EARTH SCIENCE

Internet Investigations Guide
Teacher's Edition

TERC

NATIONAL · SCIENCE · FOUNDATION

McDougal Littell

A HOUGHTON MIFFLIN COMPANY

Evanston, Illinois • Boston • Dallas

Acknowledgements

This Internet Investigations Guide, along with the corresponding Web site, was created by the Center for Earth and Space Science Education at TERC, Inc, Cambridge, Massachusetts. Funded in part by a grant from the National Science Foundation.

Project Director: Daniel Barstow

Curriculum Director: LuAnn Dahlman

Technology Director: Jennifer Loomis

Senior Scientist: Tamara Shapiro Ledley

Curriculum Developers	Technology Developers	Editorial and Production
Bryan Aivazian	Lenni Armstrong	Wesley Fleming
Brian Conroy	David Barstow	Jamey Frank
Martos Hoffman	Bryce Flynn	Tom Grace
Larry Kendall	Andree Growney	Jennifer Koh
Carla McAuliffe	Jamie Larsen	Stacey Leibowitz
Matthew Nyman	Douglas McCarroll	Jeff Lockwood
Gilly Puttick	Catherine Musinsky	Matthew Mayerchak
Zach Smith	Randy Russell	Tasha Morris
Donna Young		Sandra Schafer
		Laura Uhl
		Harvey Yazijian

Scientist Reviewers: Karl Koenig, Gerald North, Patricia Reiff, Mary Jo Richardson, David Sparks

Teacher Reviewers: Gretchen Clements, Elisa D'Amore, Howard Dimmick, Anita Honkonen, Libby Howell, Steve Jackson, Jane McMurrich, Ankie Meuwissen, Charles Mixer, Caroline Singler, Tom Vaughn

This material is based upon work supported by the National Science Foundation under grant no. ESI0095684. Any opinions, findings, conclusions or recommendations expressed in this material are those of the authors and do not necessarily reflect views of the National Science Foundation.

ISBN: 0-618-19222-0

1 2 3 4 5 6 7 8 9 BHV 05 04 03 02 01

Table of Contents

Table of Contents

UNIT 5   Atmosphere and Weather

UNIT 6   Earth's Oceans

UNIT 7   Space

UNIT 8   Earth's History

Table of Contents

Getting Started with Internet Investigations

The Internet is an ideal way to explore Earth and space science.

Through a wealth of visualizations of Earth from space, the Internet lets your students become virtual astronauts. Looking down at Earth, they see the rugged mountains of the Himalayas, peaceful island atolls, the glorious Grand Canyon, the rain forests of South America, and the vast Saharan desert.

Yet students don't just see Earth's features in isolation. They see the images and animations in a larger context, and begin to understand how the features came to be. They see the meandering Colorado River and speculate how it carved through so many layers of rocks over millions of years. They see how the Himalayas were formed as the entire Indian sub-continent collides with Nepal and China, uplifting the land into the world's highest mountains.

The Internet also helps students see how Earth's forces affect their daily lives. With satellite images of clouds and radar images of precipitation, they watch as storms build up over several hours or days, and unleash a flurry of lightning or large amounts of rain or snow. Students access Earth news and learn of volcanoes actively erupting in Hawaii, Iceland or the Philippines. When an earthquake strikes in California, they find its exact epicenter and the scope of its impact.

And when students turn their eyes to space, the Internet provides the latest images from ground-based telescopes around the world, the Hubble Space Telescope and even missions to other worlds. Students see distant galaxies, nearby stars and comets passing by Earth. When a robotic spacecraft lands on Mars, students see images from its cameras at the same time as the scientists do. And when a meteorite shower is expected on Earth, they get news well in advance so they can see these spectacular streaks of light in the nighttime sky.

Mt. Everest

NASA

Face of the Earth, ARC Science Simulations Copyright ©2001, www.arcscience.com

Indian Plate collides with Eurasian Plate, forming Himalayas

What Are Internet Investigations?

Earth Science Internet Investigations is a set of 79 Web-based investigations developed to complement the McDougal Littell *Earth Science* textbook. The investigations use satellite images, astronaut photos, interactive models, and other visuals, to help your students learn about Earth and space. The visualizations enable students to see the processes of Earth science at work. The interactivity engages students so that they are actively working with the resources, thus strengthening the learning.

Topics for the investigations are integrated with the *Earth Science, Pupil's Edition*. Each chapter has up to three investigations that are linked directly to the content in the textbook. A brief description of each investigation appears in the textbook, along with the investigation's unique keycode.

In addition to these structured investigations, the Web site includes other resources to support more open-ended student investigations:

- **Visualizations** images, animations, and models
- **Data centers** links to external resource sites
- **Earth news** links to current events relating to Earth and space
- **Local resources** maps, images, and other data for your local region
- **Careers** descriptions of careers in various fields of Earth science

Teacher's Guide

This Teacher's Guide has two sections:

1. Investigation Descriptions Each investigation is summarized in one page. Scan these pages for a quick overview of each investigation and to determine which investigations meet your needs.

2. Answered Student Worksheets Each investigation includes a student worksheet, with questions keyed directly to the Web investigation. The Teacher's Guide has answered worksheets. Students have worksheets without answers in their own student Internet Investigations Guide.

Getting Started

1. Use a computer connected to the Internet You and your students can use a computer at school, in a computer lab, at home, or wherever is most convenient. Several students can log on simultaneously. At the time of publication, the minimum computer requirements are:

Computer:	Macintosh or PC
Operating System:	Mac OS 8 or Windows 95
Browser:	Internet Explorer 4.x or Netscape version 4.x
Plug-ins:	Shockwave 7, QuickTime 4, and Flash 4

2. Connect to ClassZone.com Using your Internet browser, go to McDougal Littell's education Web site: ClassZone.com.

3. Go to the Earth Science section In ClassZone.com, select the *Earth Science* portion of the Web site. Click on *Exploring Earth* to access the investigations.

Finding Your Way around the Web Site

INVESTIGATIONS
CLASSZONE.COM

Is It Safe to Live Near a Volcano? Analyze what happened at Mount St. Helens to understand volcanoes in the Northwest United States. Develop an evacuation plan for the area surrounding Mount Rainier.

Keycode: ES0907

Enter the keycode to go directly to an investigation Each investigation has a keycode (e.g. ES0907), printed in the textbook under the description of the investigation. Enter this keycode into the search box to go directly to the investigation.

Investigations Visualizations Data Centers Earth News Local Resources Careers

Use the navigation buttons to explore other resources For a more open-ended exploration, use the buttons at the top of the screen or the list of chapters to find additional resources, such as visualizations, data centers, Earth news, local resources and careers.

Unit Investigations

For each of the eight units in the textbook, there is a special Unit Investigation that can also serve as a performance assessment. These Unit Investigations are similar to standard investigations, except that they are broader in scope and integrate Earth system interactions. Your students synthesize core concepts, analyze Earth as a system, and apply their investigation skills.

For each Unit Investigation, your students create a product or do a presentation that demonstrates their learning. Possible presentation formats are written reports, oral presentations, class debates or discussions, posters, multimedia presentations, screenplays, storyboards for a documentary, and museum exhibits. Your students may choose to develop their own questions to investigate or pursue one of the Sample Questions for Discovery provided on the Web pages.

Each unit investigation includes performance criteria for you to assess your students' products. Use the rubric on the next page (which applies to all investigations) along with the more detailed criteria specified in the description of each unit investigation. If students do alternative projects, you can use the rubric below or modify the assessment criteria to your students' projects.

Teacher's Guide
Getting Started

Understanding Core Concepts

Does the student product demonstrate a clear understanding of major concepts within the unit?

0	Exhibits minimal or a lack of understanding of core concepts
1	Exhibits an unsteady or incomplete understanding of core concepts
2	Exhibits an effective understanding of core concepts
3	Exhibits a strong understanding of core concepts
4	Exhibits an excellent and thorough understanding of core concepts

Understanding Earth System Interactions

Does the student product demonstrate a clear understanding of Earth's systems and interactions between them?

0	Exhibits an unclear or lack of understanding; Earth system phenomena are not addressed or are portrayed as independent incidents
1	Exhibits an unsteady or incomplete understanding; connections between some of Earth's systems are recognized and correctly portrayed
2	Exhibits an effective understanding; most statements about Earth's systems and their interactions are correct
3	Exhibits a strong understanding of Earth's four spheres and interactions among them
4	Exhibits mastery of Earth's systems concepts; interactions and feedback loops are correctly identified; knowledge is applied appropriately to past or future phenomena

Demonstrating Earth Science as a Process of Inquiry and Discovery

Does the student product indicate the student followed a line of inquiry or explored phenomena to discover scientific processes? Does the student product demonstrate a command of observation, interpretation and inference from data, construction or evaluation of hypotheses, development of conclusions?

0	Demonstrates no evidence of engagement in inquiry or discovery processes; incomplete or incorrect responses to posed questions; inquiry skills are applied minimally or not at all; scientific thinking skills not evident
1	Demonstrates little evidence of following a line of inquiry; responses to posed questions indicate limited engagement with content; weak application of inquiry skills; process of scientific thinking sometimes evident or inaccurately used
2	Demonstrates some evidence of following a line of inquiry; responses to posed questions indicate engagement with content; effective application of inquiry skills; scientific thinking skills generally evident and resulting in accurate explanations
3	Demonstrates clear evidence student has followed a line of inquiry; responses to posed questions indicate full engagement with content; strong application of inquiry skills; scientific thinking and analysis accurately applied
4	Demonstrates an obvious engagement with inquiry; responses to posed questions indicate full engagement with content; exceptional application of inquiry skills; process of scientific thinking consistently and accurately used as an analysis and prediction method

Presenting Results

Is the student product well organized? Do conclusions include clear statements supported by relevant citations, data, or images?

0	Exhibits no clear organization; significant components are missing; conclusions are unsupported; presentation is unappealing
1	Exhibits little organization; not all conclusions are supported; some data and graphics may not be relevant; presentation is unappealing
2	Exhibits effective organization; conclusions are supported but may not be presented completely; presentation is appealing
3	Exhibits clear organization; conclusions are well-supported; presentation is appealing
4	Exhibits exceptional organization; conclusions are well-supported and persuasive; presentation is creative and highly appealing

1 UNIT Investigating Earth

NASA

Investigate Earth from space

See Earth in new ways. Interpret photographs taken by astronauts and analyze images made by satellite instruments.

Ask questions and design investigations

Work with the same data and tools scientists use to investigate the Earth system.

Interact with 3-D models of landforms

Rotate models of mountains, cliffs, and other features to read topographic maps.

Unit 1
Investigating Earth

1 UNIT Investigating Earth

OTHER WEB RESOURCES

VISUALIZATIONS

Spark new ideas for investigations with images and animations, such as:

- Remotely-sensed images
- Earth's interacting spheres
- Paths of the water cycle
- One place at many scales

DATA CENTERS

Extend your investigations with current and archived data and images, such as:

- Earth from space
- Earth as a system
- Interactive Earth models
- Earth mapping

EARTH SCIENCE NEWS

Relate your investigations to current events around world, such as:

- Environmental news
- Satellite launches
- Earth science discoveries

Unit 1
Investigating Earth

How Are Earth's Spheres Interacting? ▶ ES0103

LuAnn Dahlman, TERC

KEY CONCEPTS

- The Earth system encompasses four interacting spheres: the geosphere, hydrosphere, atmosphere, and biosphere.
- The spheres are constantly interacting, passing energy and materials among one another.

KEY SKILLS

- Identifying portions of the Earth system
- Observing evidence of material and energy transfer
- Inferring interactions among spheres

ESTIMATED TIME REQUIRED

- 30 minutes Internet use
- 15 minutes deskwork or homework

Investigation Overview

Students view a variety of images at different scales and look for visible evidence of Earth's spheres interacting. They consider which spheres are represented in each image, and describe how materials or energy are passed between the spheres. Students follow the results of activity in one sphere through the other spheres to understand how a single process can ultimately affect the entire Earth system.

Instructional Context

Viewing our planet isolated in space can help students visualize Earth as a closed system. All processes within the closed system can be described in terms of interactions among different parts of the system. This investigation emphasizes recognition of Earth's spheres and the processes that move materials and energy among them. Once students master this task, encourage them to move beyond simply naming the spheres to comprehending the range of interactions each sphere has within the Earth system.

Teacher's Guide Chapter 1
Internet Investigation

How Do Interactions among Earth's Spheres Vary Regionally?

George Ericksen, USGS

Investigation Overview

Students examine pairs of images illustrating Earth sphere interactions at different locations around the world. They describe interactions among spheres for each pair, then compare interactions in their own region to those depicted in the images. Images of Earth from space, satellite data images, three-dimensional globe graphics, and photographs are all used to illustrate Earth sphere interactions.

Instructional Context

Students can grasp the concept of varying interactions by comparing extreme cases of sphere interactions at different locations on Earth. Once students see the range of interactions possible, they can characterize interactions in their own region more easily. Encourage students to write rich descriptions for all interactions, emphasizing the interconnectedness of the four spheres.

How Might a Scientist Investigate Annual Patterns of Fires?

BLM Fire and Aviation Image Library

KEY CONCEPTS

- Practical problems can be solved using a scientific approach.
- The risk of fire can be quantified by monitoring a variety of environmental conditions.

KEY SKILLS

- Analyzing satellite imagery
- Estimating the relative risk of fire based on environmental attributes
- Evaluating a hypothesis based on data

ESTIMATED TIME REQUIRED

- 40 minutes Internet use
- 20 minutes deskwork or homework

Investigation Overview

Students gather data from satellite images to test this hypothesis: Rating and averaging six physical conditions of an area produces an accurate prediction of the area's fire potential. Students rate the fire potential of five sites based on such factors as the amount of vegetation present, the "dryness" of potential fuels, and weather conditions. They average their individual ratings to come up with an overall predition for the risk of fire, then compare their predictions with a national fire danger map. Finally, students evaluate the hypothesis based on their results.

Instructional Context

The National Interagency Fire Center (NFIC) is a multi-organization agency charged with providing management and support to fire-fighting operations. In addition to supplying daily reports on existing fires and the potential for new fires, the NFIC also offers free public access to the satellite data images they use in their operations. These images provide the basis for their management decisions and give the public an opportunity to better understand the factors that go into making decisions about fire-fighting resources.

Instruct students to *rate* rather than *rank* the danger of fire at each location—make sure they don't rank order the locations in a sequence from highest to lowest fire danger. Tell them to consider the fire danger at each location independently of the other locations and assign a value from 1 to 5.

Teacher's Guide Chapter 2
Internet Investigation

How Might You Investigate Scientific Phenomena?

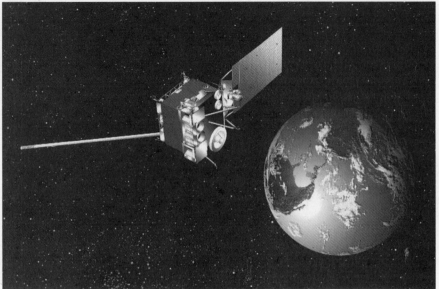

Boeing Satellite Systems

KEY CONCEPTS

- Satellite imagery depicting environmental conditions is freely available for research.
- Scientific investigations involve observing, asking questions, forming a hypothesis, designing a research method, gathering data, testing a hypothesis, and sharing results.

KEY SKILLS

- Interpreting satellite imagery
- Designing a plan to conduct a scientific investigation.

ESTIMATED TIME REQUIRED

- 45 minutes Internet use
- 15 minutes deskwork or homework

Investigation Overview

Students develop a plan for conducting a scientific investigation. They begin by exploring animations of satellite images to stimulate their interest and generate questions. Students walk through an example investigation that includes formulating a hypothesis from a question, gathering data to test the hypothesis, and drawing a conclusion about the validity of the hypothesis. Using the example as a model, students outline their own plan for conducting an investigation using images from NASA's Earth Observatory Website.

Instructional Context

Visit the NASA Earth Observatory (http://earthobservatory.nasa.gov) to preview the extraordinary variety of well-catalogued and easily accessed satellite imagery available for research projects. Note especially the capability of creating time-series animations comparing two environmental conditions in a linked animation. The highly visual nature of the data may inspire students to follow their own line of inquiry in developing an investigation. Consider implementing students' investigation plans as a follow-up project.

How Do Map Projections Distort Earth's Surface?   ▶ ES0301

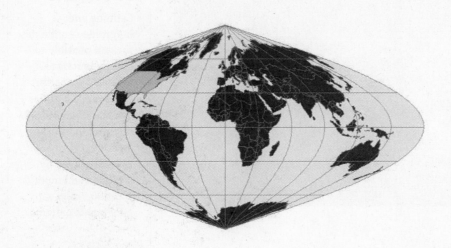

Carla McAuliffe, TERC

Copyright © McDougal Littell Inc.

KEY CONCEPTS

- Maps can display a variety of information at a range of scales.
- Maps represent a three-dimensional Earth on a two-dimensional surface.
- Map projections preserve up to two of the following: distance, area, or shape.
- Only a globe can accurately preserve distance, area, and shape.

KEY SKILLS

- Quantifying observations through measurement
- Interpreting maps

ESTIMATED TIME REQUIRED

- 45 minutes Internet use

Investigation Overview

Students begin this investigation by exploring a series of maps showing various locations at different scales. Students compare and contrast the images to compile a list of the basic properties of maps. Next, they consider how maps represent a three-dimensional Earth on a two-dimensional surface. They are introduced to map projections through a series of animations that show the Earth's surface peeling off a globe and transforming into a flat map projection. Students measure distance and areas and describe the shapes of features across three different map projections. They learn that each projection is useful for portraying specific regions of Earth.

Instructional Context

Map projections can be a difficult concept for students to understand. Working with 3-D models can make it easier for students to picture the geometric relationships between a globe and a projection surface. Have students wrap pieces of paper or clear acetate around globes to form cylinders and cones that represent projection surfaces. They should hold paper flat against the North or South Pole to simulate planar projections. For each of these models, ask students where the projection surface touches the globe. Help them recognize that the distortion of a map projection is minimal in the region where the projection surface makes contact with, or is tangent to, the globe. Distortion increases with distance from the contact point.

Teacher's Guide Chapter 3
Internet Investigation

How Do Latitude and Longitude Coordinates Help Us See Patterns on Earth?

▶ ES0303

Carla McAuliffe, TERC

KEY CONCEPTS

- Latitude and longitude coordinates make it possible to specify precise locations on Earth.
- Geographic patterns can be difficult to interpret in tabular form.
- Data displayed by latitude and longitude coordinates reveal geographic patterns.

KEY SKILLS

- Plotting latitude and longitude coordinates onto maps
- Analyzing geographic patterns
- Evaluating the relevancy of using latitude and longitude.

ESTIMATED TIME REQUIRED

- 45 minutes Internet use

Investigation Overview

Students inspect a table of sea surface temperatures recorded at specific latitude and longitude coordinates. They have difficulty discerning any relationship between location and temperature. Students review the concepts of latitude and longitude with an interactive animation. Next, they organize the data geographically by plotting the indicated temperatures according to latitude and longitude. After plotting the data, students can discern the relationship between temperature and location easily. They use colored pencils to show the pattern on their printed maps. Students observe animations of sea surface temperature over time to recognize that some natural processes vary through time as well as by location. Finally, students are introduced to global positioning system (GPS) technology as a means of gathering latitude and longitude data.

Instructional Context

Latitude and longitude are paired concepts that, when taught together, can be confusing for students. This investigation gives students the opportunity to look at each set of lines separately. They toggle latitude and longitude lines on and off a global map to comprehend or review these concepts. Latitude is a primary controlling factor of climate; lines of longitude are directly tied to methods of keeping time. A discussion of these topics could be used for either introducing or culminating this investigation.

How Are Landforms Represented on Flat Maps? ▶ ES0307

NASA/JPL/NIMA/USGS

Copyright © McDougal Littell Inc.

KEY CONCEPTS
- Topographic maps use contour lines to represent Earth's surface.
- Closely spaced contours indicate steep slopes, while widely spaced contours indicate gentle slopes.
- Hachures on contour lines show decreases in elevation.

KEY SKILLS
- Interpreting topographic maps
- Identifying landform features such as hills, valleys, and cliffs on topographic maps
- Visualizing three-dimensional landforms from two-dimensional contour lines

ESTIMATED TIME REQUIRED
- 45 minutes Internet use

Investigation Overview

Students begin by watching a simulated flyby over a river and sand dunes in the Namibia Desert of Africa. They make detailed observations about the topography they encounter during the flyby. Then, students are introduced to the topographic map, a specialized type of map that uses contour lines to show the shape of Earth's surface. Students explore animated, 3-D models of surface features to find out how contour lines represent hills, valleys, steep areas, and depressions. Students try to recognize features on topographic maps as they take a topographic tour of major landforms of the United States. Last, they discover how space-based technologies are contributing to the design of more detailed topographic maps showing remote areas of the world that have historically been difficult to map.

Instructional Context

Visualizing landforms represented on a topographic map can be a difficult skill for students to master. This spatial task requires students to translate two-dimensional lines on the map into three-dimensional images in their minds. In addition to working with computer models of topography, having students create their own topographic maps from physical models can be a very useful strategy. Students can make clay models of geographic features. If they place these models into large containers and cover them with water at one centimeter intervals, students can generate a contour map by drawing where the water meets the model. Also of great value would be taking students outside with a local topographic map of their area and helping them recognize key features on their maps.

How Can Getting Farther Away from Earth Help Us See It More Clearly?

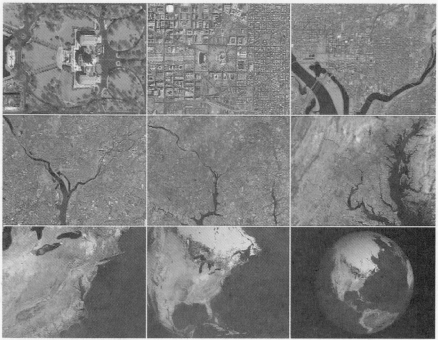

NASA/GSFC

Investigation Overview

This investigation focuses on the use of high-altitude imaging to observe the Earth system. Students examine a series of images that show a location on Earth from progressively higher altitudes. Using animations and illustrations, students explore basic principles of satellite imaging, including multispectral and time series imaging. As a culminating project, students research Earth observation satellites currently in use, noting the purpose of each and how they achieve that purpose. Finally, students plan their own Earth observation mission and prepare a presentation on the mission and its objectives.

Instructional Context

High-altitude observations are largely responsible for our understanding of Earth as a set of interconnected systems. Aircraft and satellites help us overcome limitations inherent in our ground-based observational capabilities by extending our abilities to see larger areas at many wavelengths over time. It would be helpful for students to have a basic understanding of the electromagnetic spectrum prior to beginning this investigation.

KEY CONCEPTS

- High-altitude remote sensing extends our ability to see Earth.
- Moving farther away from Earth provides a large-scale view of the Earth system.
- Individual satellites record unique types of information.

KEY SKILLS

- Interpreting aerial and satellite images
- Explaining the advantages and disadvantages of satellite remote sensing
- Planning a simulated Earth-observation satellite mission
- Communicating information about the student-designed mission

ESTIMATED TIME REQUIRED

- 90 minutes Internet use
- 60 minutes deskwork or homework

How Can Getting Farther Away from Earth Help Us See It More Clearly?

Unit 1 Project Description

Using the results of this investigation and Web-based research, students design their own Earth observation satellite mission. Students prepare a presentation that describes their satellite, its mission, and how it functions. Students build a case for supporting their mission based on data and research. They present their findings in a format that you have approved.

Criteria for setting student expectations and grading the project are provided below. In conjunction with the criteria, you may also want to use the assessment rubric to evaluate student projects. The rubric is found in the Introduction of the Teacher's Guide.

Unit Project Criteria

1. **A description of the proposed satellite mission. (20%)**
   - Name the satellite
   - Describe the purpose of the mission
2. **An explanation of how the satellite functions. (30%)**
   - Describe the spectral characteristics (wavelength of radiation that it will use)
   - Specify the data collection method (time series images of one area, or just one image of many places)
   - Describe the type of orbit (polar or geostationary)
3. **A case for supporting your mission. (35%)**
   - Provide examples of the kinds of questions you hope to answer using data the mission collects
   - Establish a need for the unique data that your mission will collect
   - Give a rationale for why this satellite should be built
4. **An effective presentation of your conclusions. (15%)**
   - Communicate findings and ideas clearly and accurately
   - Present appropriate data that support your conclusions
   - Use appealing visuals and design

Alternative Projects

Students may want to choose their own research topic or pursue one of the Questions for Discovery suggested below. The assessment rubric in the Teacher's Guide provides a standard basis for evaluating student performance and is recommended for grading alternative projects.

Sample Questions for Discovery

- How have Earth-observing satellites changed with advancing technology?
- How does the Landsat 7 mission compare to the SIR-C mission?
- What new Earth-observing systems are planned for the future?

2 UNIT Earth's Matter

Face of the Earth™ ARC Science Simulations Copyright © 2001
www.arcscience.com

Explore how rocks change over millions of years

Read the clues in a rock to tell how it formed. Follow it through the rock cycle to see what it might become next.

Choose between paper or plastic

Design and conduct your own research project to decide which type of bag is best for the planet. Create an informational product to convince others of your choice.

Track an oil spill

Investigate how ocean currents and winds move oil spills. Learn how humans try to minimize the damage from these environmental accidents.

2 UNIT Earth's Matter

OTHER WEB RESOURCES

VISUALIZATIONS

Spark new ideas for investigations with images and animations, such as:

- Origin of the Solar System
- Earth cross-sections
- Rocks under a microscope
- 3-D models of molecules

DATA CENTERS

Extend your investigations with current and archived data and images, such as:

- Rocks and minerals
- Methods of keeping time
- Solar Power
- National Parks

EARTH SCIENCE NEWS

Relate your investigations to current events around world, such as:

- Environmental news
- Satellite launches
- Earth science discoveries

Copyright © McDougal Littell Inc.

Unit 2
Earth's Matter

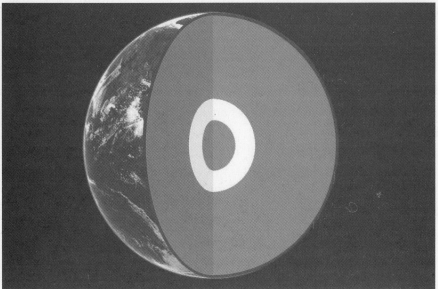

KEY CONCEPTS

- The interior structure of Earth can be inferred from the trajectories of P and S waves moving through the planet.
- Models can be used to simulate conditions and conduct experiments when direct observation is not possible.

KEY SKILLS

- Interpreting graphs
- Visualizing earthquake waves moving through Earth's interior
- Modeling the interior structure of planets

ESTIMATED TIME REQUIRED

- 45 minutes Internet use

Investigation Overview

Students begin the investigation by considering the dynamic nature of Earth both above and below its surface. Next, they explore earthquake waves and their paths through different materials. Students view a series of animations showing P and S waves traveling through model planets with different interior compositions. They observe the P and S wave behavior in each model and compare the behavior to the structure of the planet. After exploring these animations, students examine P and S wave data for Earth and use the information to estimate the depth and state of Earth's layers. Last, students investigate seismic tomography models that are helping to refine three-dimensional views of Earth's structure.

Instructional Context

Recent advances in seismic technology allow researchers to map areas where seismic waves move anomalously fast or slowly. This research, referred to as seismic tomography, facilitates development of three-dimensional models that are useful for studying the details of plate tectonic processes. While these models are not fictional, they are theoretical and mathematically derived. Help students understand that scientists use earthquake waves as an inferential tool. Researchers try to build models that approximate reality. Discuss with students how models help scientific researchers simulate conditions and run experiments in ways that would otherwise not be possible.

What Time Is It?

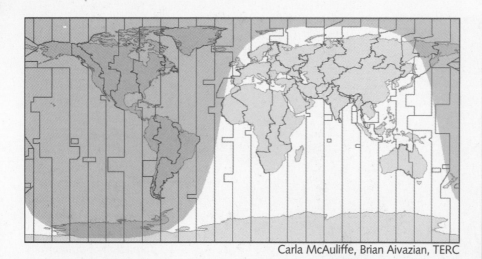

Carla McAuliffe, Brian Aivazian, TERC

KEY CONCEPTS

- Day and night occur because Earth rotates.
- The sun is always rising and setting somewhere on Earth.
- Earth is divided into twenty-four time zones, each spanning approximately fifteen degrees of longitude.

KEY SKILLS

- Relating Earth's rotation to concepts of time
- Interpreting maps
- Determining longitude from time and vice versa

ESTIMATED TIME REQUIRED

- 45 minutes Internet use

Investigation Overview

Students begin this investigation by exploring the concepts of day and night. They examine still images of Earth at a single moment in time. By interacting with animated models of a rotating Earth, students discover that sunrise and sunset are always happening somewhere on the planet. Next, students consider how time relates to the apparent motion of the sun. They investigate "suntime" using animations of geochrons and sundials. After being introduced to time zones, 15 degree intervals of longitude that share the same time, students interact with time zone models to figure out what time it is in different parts of the world. Finally, in a manner similar to methods used by navigators, they rely on time to locate their position on Earth.

Instructional Context

The apparent motion of the sun is a source of confusion for people of all ages. Help students recognize that if Earth did not rotate, a day would be as long as a year, with six months spent in darkness and six months in daylight. Kinesthetic and physical models of Earth rotating and revolving around the sun can be useful aids for promoting student understanding of time-related processes. Have students stand in a circle with a model sun in the center, and tell them that their heads represent Earth. As they rotate to the left, students can correlate sunrise, noon, sunset, and midnight with specific positions in Earth's daily rotation. The time the sun rises and sets each day is influenced not only by Earth's rotation, but also by seasonal changes as our planet revolves around the sun. The seasonal complexity is not addressed in this investigation, but it is modeled in Visualization ES0408.

How Many Protons, Neutrons, and Electrons Are in Common Elements?

Zach Smith, TERC compiled from Arttoday.com

Investigation Overview

Students manipulate the number of subatomic particles (protons, neutrons, and electrons) in a simplified model to build atoms of common elements. Students add subatomic particles to the model to construct atoms, isotopes, and ions (unbalanced atoms). They receive instant feedback on the element's name and any additional particles needed to make it stable and balanced. Students use the model to build atoms of simple elements. Then, they use a periodic chart to determine the number of subatomic particles in several heavier elements.

Instructional Context

The small size of atoms precludes direct visualization of their internal structure and composition. This interactive computer simulation gives students the opportunity to examine visual models that represent abstract items. Though the model is very simple, it is useful for preparing students to comprehend how chemical compositions affect the properties of minerals.

KEY CONCEPTS

- The number of protons in an atom determines the element.
- The number of protons plus neutrons determines the mass of an atom.
- Varying the number of neutrons in the nucleus of an atom results in the formation of different isotopes of that element.
- When an atom has unequal numbers of protons and electrons, its charge is unbalanced and it is called an ion.

KEY SKILLS

- Manipulating (in a simulation environment) subatomic particles
- Analyzing the behavior of subatomic particles in a model

ESTIMATED TIME REQUIRED

- 35 minutes Internet use

How Do Crystals Grow?

Smithsonian Institution

KEY CONCEPTS

- Crystals grow as atomic particles are added to their ordered crystal structure.
- Temperature, pressure, and space limitations affect the growth of crystals

KEY SKILLS

- Observing crystal growth
- Predicting crystal shapes
- Analyzing the effects of changing conditions on crystal growth

ESTIMATED TIME REQUIRED

- 35 minutes Internet use

Investigation Overview

Students observe time-lapse video clips of crystals growing. They examine crystal forms and relate them to the structure of the atoms that make up the crystals. Students rotate three-dimensional lattices of crystals to view them from several perspectives. Students consider what variables might affect crystal growth, and in an interactive simulation, they select conditions under which to model crystal growth and view the results.

Instructional Context

Crystal growth occurs at the atomic level, and often takes so long that the process is difficult to observe. Time-lapse video clips enable students to witness crystal growth at an accelerated pace and at a size that allows them to analyze how crystals grow. This investigation emphasizes growth of crystals from aqueous solutions rather than magmatic crystal growth. Students can grow crystals under a microscope in class: Prepare saturated solutions of compounds such as $NaCl$, $CuSO_4$, $MgSO_4$, and $CuCl_2$ in small test tubes with stoppers. Apply a small amount of solution to a microscope slide by touching the bottom of the wet test tube stopper to the slide. As students watch through the microscope, the heat of the lamp will cause the water to evaporate, and crystals will form.

How Do Rocks Undergo Change?

Matthew Nyman, TERC

KEY CONCEPTS

- Rock-forming processes move rocks through the rock cycle.
- Rocks are classified as igneous, sedimentary, or metamorphic based on how they formed.

KEY SKILLS

- Interpreting diagrams
- Observing and summarizing igneous, sedimentary, and metamorphic processes

ESTIMATED TIME REQUIRED

- 30 minutes Internet use

Investigation Overview

Students follow a hypothetical rock through the rock cycle on an interactive diagram. Each arrow in the diagram reveals an animation of the rock-forming processes it represents. Students observe that each of the three rock types can be transformed into other rocks. Students then write detailed descriptions of how rocks change to form igneous, sedimentary, and metamorphic rocks.

Instructional Context

You can introduce the rock cycle by comparing it to the familiar concept of the water cycle. Draw a rough outline of the water cycle and ask students to describe the processes that move water from one part of the cycle to another. Present the rock cycle by explaining that rock-forming processes move rocks from one part of the rock cycle to another, continually forming the three types of rocks. You may want to discuss the disruptive effects of human activities on the rock cycle. Ask students to consider how human processes such as road building and melting ore mimic natural rock-cycle processes.

How Do Igneous Rocks Form?

Karl Karlstom

KEY CONCEPTS

• Crystal size in igneous rocks reflects cooling rates of the magma.

KEY SKILLS

• Interpreting images
• Observing differences in rock texture
• Inferring cooling rates from rock texture

ESTIMATED TIME REQUIRED

• 35 minutes Internet use

Investigation Overview

Students view and describe images showing clear examples of coarse-grained, fine-grained, and porphyritic igneous rock textures. The term *igneous rock texture* is introduced; students learn that the igneous rock texture can be used to infer cooling rates of magma. Students view animations illustrating formation of the three textures. Students examine photographic images of igneous rocks and use their knowledge to draw conclusions about the cooling rate of the magma that formed the rocks.

Instructional Context

You may want to model the process of igneous rock formation with this simple demonstration: Melt three thumb-sized chunks of sulfur in a test tube over a propane torch or bunsen burner. This molten liquid is your model magma. Cool a portion of the sulfur very slowly (pour it into a warmed crucible). Cool some of the sulfur somewhat more quickly (pour a portion into a depression in a piece of aluminum foil). Cool the last portion of the sample very quickly (pour into a shallow pan of water). Have students use a hand lens to view the model igneous rocks that were formed at different cooling rates.

Teacher's Guide Chapter 6
Internet Investigation

What Kind of Rock Is This?

LuAnn Dahlman, TERC

KEY CONCEPTS

• Rocks can be identified by their composition and texture.

KEY SKILLS

• Observing rock textures and compositions
• Identifying rock textures and compositions

ESTIMATED TIME REQUIRED

• 30 minutes Internet use

Investigation Overview

Students follow an interactive Web-based key to identify rock samples provided by the instructor. The key is coupled with images of rocks at scales ranging from outcrops to hand specimens. Rock types addressed in the key include: *Igneous*: Granite, Rhyolite, Diorite, Andesite, Gabbro, Basalt, Obsidian, Pumice, and Scoria. *Sedimentary*: Sandstone, Shale, Conglomerate, Limestone, and Coal; *Metamorphic*: Quartzite, Marble, Slate, Phyllite, Schist, and Gneiss

Instructional Context

Use the Web-based key to identify rock samples available in your classroom. You may want to work with a group of samples that can be correctly identified with the key. Affix sample numbers to these and offer them as the unknown samples for students to classify. The clearest examples of most rock types should be easy to identify using the key. However, a substantial percentage of rocks may not have enough distinguishing features to be identified with the key.

What Happens When an Oil Spill Occurs? ▶ ES0703

U.S. Coast Guard

Investigation Overview

Students use the case of the oil spill from the *Exxon Valdez* to explore what happens when an oil spill occurs. Students examine images taken during the spill, then follow the spill response effort through the steps of containment, modeling, and impact assessment. Next, students work with an oil spill simulation to discover how wind speed and direction affect the trajectory of an oil spill. Last, students predict which locations along the shore would be affected by a simulated oil spill.

Instructional Context

Oil spill response efforts rely on detailed and accurate observations. Because many factors affect the trajectory of oil spills, they are monitored using computer models. The models allow the response team to predict where the oil will move next. Models are adjusted using detailed observation notes gathered on-site and during flights over the oil spill area. Impress upon students that the skills necessary for successful response to an oil spill (observing, interpreting, predicting) are the same ones inherent in the scientific method.

KEY CONCEPTS

- Responding to an oil spill requires observation, prediction, and analysis.
- Aerial photography and computer models are used to track the flow of spilled oil.
- Wind speed, wind direction, flow of the current, and composition of the oil each contribute to the movement of an oil spill.

KEY SKILLS

- Modeling oil spill trajectories
- Predicting environmental impacts
- Analyzing environmental conditions

ESTIMATED TIME REQUIRED

- 45 minutes Internet use

Teacher's Guide Chapter 7
Internet Investigation

Why Is This Place Protected?   ▶ ES0705

NPS

KEY CONCEPTS

- The wise use of natural resources enhances our lives.
- Natural resources may be either renewable or nonrenewable.
- National parks, monuments, and preserves exist to protect natural resources and wonders.

KEY SKILLS

- Identifying natural resources present in national parks, monuments, and preserves.
- Categorizing natural resources as renewable or nonrenewable.
- Evaluating unprotected areas to determine if they are worth protecting.

ESTIMATED TIME REQUIRED

- 45 minutes Internet use
- 45 minutes deskwork or homework

Investigation Overview

As the population of the United States continues to grow, pristine habitats face increasing pressures from development. One method of preserving natural resources is through the establishment of national parks, monuments, and preserves. Students take virtual field trips to protected places and describe the unique features they find. They classify the natural resources present in protected areas as either renewable or nonrenewable. Finally, they use the criteria for establishment of a national park to determine if another place is worth protecting.

Instructional Context

To motivate student interest in preservation issues, locate the nearest protected and unprotected areas closest to your school. (A city park and a construction site could serve as good local examples of areas that are and are not protected.) Consider taking students on field trips to both areas. Allow them to compare the unique features of each place. Many environmental organizations play a role in preserving natural resources. Consider inviting someone from one of these agencies into your classroom as a guest speaker.

What Environmental Changes Can We See with Satellites?

NASA, Goddard Space Flight Center Scientific Visualization Studio

KEY CONCEPTS

- Satellites and other space-based imaging technologies provide a large-scale perspective for monitoring environmental changes.
- The rate of many environmental changes is exponential, not linear.

KEY SKILLS

- Interpreting satellite images
- Quantifying observations through measurement
- Analyzing environmental changes

ESTIMATED TIME REQUIRED

- 45 minutes Internet use

Investigation Overview

Students explore a series of satellite images of the Las Vegas, Nevada metropolitan area. The images, taken in 1964, 1972, 1986, and 1992, show the increasing extent of urban development. Students use animation to help them visualize growth patterns. They quantify their observations by measuring the area of development through time. Students consider the factors that play a role in urban development, and they characterize and analyze growth patterns by plotting city size and population as a function of time. After drawing conclusions about urban growth, students explore additional satellite images to investigate other cases of environmental change over time.

Instructional Context

Satellite images offer large-scale views of Earth. Natural and human-induced changes in the environment are visible in many satellite images. Changes can be recognized by comparing images over time, or by looking for disruptions (usually geometric, like straight lines) in the landscape. Satellites record many types of images, from visible light photographs (essentially a color photo) through false color images that depict infrared and other wavelengths of light. Some images will be very familiar to students. Others, such as false color composites, will be less recognizable. Help students discern what the images are showing.

Teacher's Guide Chapter 7
Internet Investigation

Paper or Plastic–Which Type of Bag Is Better for the Environment?

LuAnn Dahlman, TERC

Teacher's Guide Unit 2
Internet Investigation

KEY CONCEPTS

- The Earth system has limited resources.
- Issues facing society are multi-faceted.
- Credible cases can be made for made for more than one sid eof an issue.

KEY SKILLS

- Analyzing the range of complexity presented by a single issue
- Researching aspects of a complex issue
- Presenting a balanced argument and making a credible case for one side

ESTIMATED TIME REQUIRED

- 60 minutes Internet use
- 90 minutes deskwork or homework

Investigation Overview

Students are provided with a procedural framework for research, presented in the context of deciding if paper or plastic bags are better for our planet. Five steps for conducting research are outlined, each with examples of how to accomplish that step for the investigation at hand. Students will select one or more factors affecting the paper-versus-plastic issue. For each, they do basic research using the Internet or print-based resources, or by conducting interviews or experiments. Their findings will form the basis of a balanced argument, and will conclude with students taking a position in favor of one of the options. Students communicate the results of their research in a teacher-approved format.

Instructional Context

Problems rarely have a single dimension. More often, problems can be viewed from a number of perspectives. This investigation requires students to consider the paper-versus-plastic question from many perspectives and to make a decision based on their own findings. The Internet links offered as part of this investigation expose students to a variety of data sources, many of them with very specific self-interests. Discuss ways to evaluate the validity of information presented on the Internet and in other media.

Paper or Plastic–Which Type of Bag Is Better for the Environment?

Unit 2 Project Description

How does a choice as simple as deciding which type of bag to use at the grocery store affect the larger Earth system? Using Web resources, students consider the paper-versus-plastic question from many perspectives and make a decision based on their findings. They select one or more factors relating to the decision of which type of bag to choose, then use the information to present a balanced argument in favor of one of the options. Students present their findings in a format that you have approved.

Criteria for setting student expectations and grading the project are provided below. In conjunction with the criteria, you may also want to use the assessment rubric to evaluate student projects. The rubric is found in the Introduction of the Teacher's Guide.

Unit Project Criteria

1. A comprehensive list of issues related to the question (10%)
- Develop an extensive list of issues related to the manufacture, use, and disposal of paper and plastic bags.
- Include ideas that come from a variety of different categories.

2. Evidence of credible research (30%)
- Demonstrate that a number of different sources were used for research.
- Show that a variety of types of resources were consulted.
- Present evidence of research in the form of useful notes.
- Make appropriate references to the sources you consulted.

3. A balanced argument (20%)
- Provide information on both types of bags in a balanced manner.
- Identify potential biases in your sources of information.

4. A credible case for your position (20%)
- Take a stand on one side of the issue.
- Defend your position logically and support your arguments with citations or data.

5. A presentation of your research findings (20%)
- Communicate findings and ideas clearly and accurately.
- Present appropriate data.

Alternative Projects

Students may want to choose their own research topic or pursue one of the Questions for Discovery suggested below. The assessment rubric in the Teacher's Guide provides a standard basis for evaluating student performance and is recommended for grading alternative projects.

Sample Questions for Discovery

- How do social, economic, and environmental perspectives come into play when potential impacts of any modern problem are considered?
- How might small, daily personal choices affect the larger global system?
- What are the long-term environmental implications of using disposable products?

3 UNIT Dynamic Earth

NASA

Check out active volcanoes
Access volcano-cams to look for current activity on volcanoes around the world.

Investigate earthquakes in Los Angeles
Analyze ground motion around Los Angeles to pinpoint the location of a blind thrust fault.

Explore mountain belts
Check out some of Earth's most prominent mountain chains. Examine photos of geological features from each range to figure out how the mountains formed.

Unit 3
Dynamic Earth

Earth Science

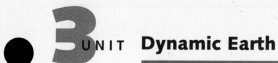

UNIT Dynamic Earth

OTHER WEB RESOURCES

VISUALIZATIONS

Spark new ideas for investigations with images and animations, such as:

• Motion of the plates

• Different types of volcanoes

• Earthquake videos

• Mountains forming

DATA CENTERS

Extend your investigations with current and archived data and images, such as:

• Live volcano-cams

• Earthquake seismic data

• Safety tips

• Volcanoes on other worlds

EARTH SCIENCE NEWS

Relate your investigations to current events around world, such as:

• Active volcanic eruptions

• Earthquakes around the world

• New discoveries about tectonics

Unit 3
Dynamic Earth

What Is Earth's Crust Like? ▶ ES0801

NASA

KEY CONCEPTS

- Volcanoes are evidence of Earth's internal heat escaping at the surface.
- Earthquakes occur where solid rocks are in motion relative to other solid rocks.
- Folded mountain belts reflect absorption of energy applied perpendicular to their length.
- The Earth's crust is composed of rigid plates. Geologic processes and features are concentrated along plate boundaries.

KEY SKILLS

- Interpreting maps
- Inferring internal processes from external events

ESTIMATED TIME REQUIRED

- 30 minutes Internet use

Investigation Overview

Students explore a map showing the rocky surface of Earth including topography of the seafloor. They turn on overlays of data showing the locations of recent volcanoes, earthquakes, and folded mountain belts. Students discover that our planet's internal heat and energy are released (or absorbed in the case of mountain building) in these locations. Students infer that volcanoes, earthquakes, and mountain belts reveal zones of weakness (eventually identified as plate boundaries) in Earth's outer shell.

Instructional Context

This investigation helps students build a mental model of Earth's outer "shell" by focusing on observable geologic processes. You might follow this investigation by using a simple hard-boiled egg with cracks in the shell as a model for Earth. Completing this investigation before presenting information about "continental drift" may help students avoid the misconception that continents, rather than plates, move across Earth's surface.

How Old Is the Atlantic Ocean?

NOAA

KEY CONCEPTS

- The age of an ocean that split two continents can be inferred from the oldest rocks on its seafloor.
- The rate of seafloor spreading can be calculated using age of the seafloor and distance between the continents.

KEY SKILLS

- Interpreting maps and map symbols
- Inferring the age of an ocean from age of seafloor rocks
- Calculating rate from distance and time
- Hypothesizing about former continental arrangements

ESTIMATED TIME REQUIRED

- 35 minutes Internet use

Investigation Overview

Students observe the match of South America's and Africa's coastlines, then examine an image modeling the two continents reconstructed as one. They examine evidence that supports the idea that the two continents were once joined, then explore an image showing topography of the Atlantic seafloor. Students then examine patterns showing the age of rocks on the seafloor and infer that the oldest rocks indicate the age of the Atlantic Ocean. Based on the current distance between the continents, and the amount of time they have been moving apart, students calculate the average rate of seafloor spreading in the Atlantic Ocean.

Instructional Context

By the time they reach high school, many students are already familiar with the theory that separate continents were once joined. This investigation allows them to use that understanding along with new data to date the opening of the Atlantic Ocean. You can extend the investigation by having students draw reconstructions for continents on opposite sides of a mid-ocean ridge.

Teacher's Guide Chapter 8
Internet Investigation

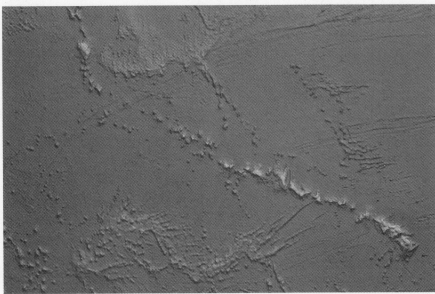

Jennifer Loomis, TERC

KEY CONCEPTS

- Hotspots are revealed by volcanic activity that is not associated with a plate boundary.
- The rate of plate motion over a hotspot can be calculated using the age of volcanic rocks produced over the spot and their distance from the hotspot.

KEY SKILLS

- Interpreting maps
- Hypothesizing about island formation
- Measuring distances on maps
- Calculating rate of motion

ESTIMATED TIME REQUIRED

- 45 minutes Internet use

Investigation Overview

Students explore a relief map and images of the Hawaiian Islands to predict their relative ages. After students examine the ages of volcanic rocks on each island, they are asked to suggest a model of island formation that explains why the age of islands increases with distance from the big island of Hawaii. Students view an animation of progressively younger volcanic islands forming over a hotspot as older volcanoes are moved farther from the hotspot by plate motion. Students measure the distance islands have moved from the current location of the hotspot and use the age of volcanic rocks to calculate the rate of the Pacific plate's motion. Finally, students discern the direction and calculate the rate of the North American plate's motion over the hotspot currently located under Yellowstone National Park.

Instructional Context

When students first consider the volcanoes of Hawaii, far from any plate boundaries, they may see the volcanoes as evidence that does not fit with the theory of plate tectonics. Students will discover how hotspot volcanism supports plate tectonics theory and that it provides a simple method to quantify the rate of plate motion. Emphasize that hotspots are located within the mantle; hotspots remain relatively stationary while the lithosphere moves over them. Ask students to consider how hotspot traces might help in reconstructing past arrangements of the continents.

How Are Volcanoes Related to Plate Tectonics?   ▶ ES0901

Jennifer Loomis, TERC

KEY CONCEPTS

- Rift zone volcanism occurs directly along plate boundaries
- Subduction-related volcanoes occur at a distance from the plate boundary along a line parallel to their subduction zone.
- Processes at convergent and divergent boundaries produce distinctly different types of volcanism.

KEY SKILLS

- Plotting volcanoes by latitude and longitude
- Correlating map views with subsurface views
- Inferring plate tectonic processes from map views

ESTIMATED TIME REQUIRED

- 45 minutes Internet use

Investigation Overview

Students use latitude and longitude coordinates to plot volcano locations on a relief map of the Pacific Northwest. Based on the shape of the land and the locations of volcanoes, students predict the locations of plate boundaries in the region. Students refine their predictions after they see the shape of the sea floor. When they see the actual plate boundary locations, students try to infer the subsurface processes that produce volcanoes on the surface. A 3-D block diagram shows the relationship between the map view and subsurface geology and illustrates the fundamental difference between rift zone and subduction-related volcanism. Finally, students examine maps of other plate boundaries and associated volcanism and infer how the volcanoes are related to plate tectonic processes.

Instructional Context

Based on simplified models of plate tectonics, students may believe that any line of volcanoes indicates the presence of a plate boundary. When asked for predictions, students will typically draw plate boundaries along both lines of volcanoes on the map. Once the subsurface geology is revealed, students see that plate tectonic processes produce two fundamentally different types of volcanism. Before students start the investigation, review a simple model illustrating the relative motions possible on either side of a plate boundary (use two hands or books to model the motion). As a follow-up to the investigation, provide students with access to 3-D animations or physical models to help them develop the skill of correlating map views with subsurface processes.

How Fast Do Gases from Volcanic Eruptions Travel?

▶ **ES0906**

James Valance, USGS, CVO

KEY CONCEPTS

- Explosive volcanic eruptions loft gases and ash into the atmosphere.
- Satellite images taken at recorded times provide a way to calculate rate of motion of volcanic ash and gases.
- Aerosols from volcanic eruptions remain suspended in the atmosphere for many months.

KEY SKILLS

- Identifying volcanic materials in satellite images
- Measuring distances
- Calculating rates of motion

ESTIMATED TIME REQUIRED

- 45 minutes Internet use

Investigation Overview

Students examine animations of satellite images taken following the June 1991 eruption of Mount Pinatubo in the Philippines. They identify volcanic materials in the images and track the location of the materials over time. Students measure the distance volcanic gas and ash particles (aerosols) traveled over five days and use the data to calculate the rates of motion. They also calculate the amount of time it would take aerosols from the eruption to circle the globe, and they consider the effect of increased aerosols on Earth's climate.

Instructional Context

The state of our planet is well monitored by satellites. Weather satellites record updated views of Earth several times per day, and NASA's Total Ozone Mapping Spectrometer (TOMS) monitors concentrations of gases and aerosols in the atmosphere. Animated satellite images offer a dynamic view of volcanic ash and gases spreading away from the Pinatubo eruption site and around the globe. It is essential for students to understand how concentration of a substance differs from its total amount. In the case of Mount Pinatubo, high concentrations of sulfur dioxide and aerosols decreased within a few weeks of the eruption. Total amounts of these volcanic products in the atmosphere decreased much more slowly.

Is It Safe to Live Near a Volcano?

David E. Wieprecht, USGS

Investigation Overview

Students peruse interactive Web pages detailing the devastation caused by the 1980 eruption of Mount St. Helens. They examine historical eruption frequencies of Cascade volcanoes and consider the possibility of another one erupting in their lifetime. Because areas of dense population are near Mount Rainier, students consider the steps that would need to be taken to save lives if it were to erupt in a manner similar to Mount St. Helens. Students examine maps showing the extent of historical mudflows from Mount Rainier and delineate areas they think would be in danger if it were to erupt again. Finally, they outline an evacuation plan that could be implemented if an eruption of Mount Rainier seemed imminent.

Instructional Context

Cascade volcanoes will probably continue erupting in a manner similar to the 1980 eruption of Mount St. Helens. The area of devastation from such eruptions can extend far beyond the volcanic mountain itself. Mudflows race down existing streams and rivers, burying everything in their path. The blast of hot volcanic gas can also devastate living things in a large region. People who live in or visit areas near Cascade volcanoes should make themselves aware of evacuation plans in effect for the region.

KEY CONCEPTS

- Eruptions from subduction-related volcanoes endanger life in a large region surrounding the volcanic vent.
- Maps of historical mudflows help delineate areas vulnerable to future eruptions.
- Evacuation planners must consider a range of issues to ensure public safety.

KEY SKILLS

- Interpreting maps and 3-D topographic models
- Describing volcanic damage
- Designing an evacuation plan for a volcanic area

ESTIMATED TIME REQUIRED

- 45 minutes Internet use
- 20 minutes deskwork or homework

Teacher's Guide Chapter 9
Internet Investigation

How Are Earthquakes Related to Plate Tectonics?

Larry Kendall/The Living Earth

Copyright © McDougal Littell Inc.

KEY CONCEPTS

- Most earthquakes occur along or near plate boundaries.
- Most earthquakes are small and occur at shallow levels in the Earth's crust.
- Convergent plate boundaries experience the largest and deepest earthquakes.

KEY SKILLS

- Interpreting map patterns
- Hypothesizing about types of earthquakes at different plate boundaries
- Analyzing susceptibility to earthquakes

ESTIMATED TIME REQUIRED

- 30 minutes Internet use

Investigation Overview

In this investigation, students examine different overlays on a world relief map to interpret the distribution of earthquakes. Students look at the numbers of shallow, intermediate, and deep-focus earthquakes and conclude that subduction zones are the type of plate boundary where deep-focus earthquakes occur. Students then compare earthquakes of different magnitudes, determining that small-magnitude earthquakes are the most abundant, occurring along all types of plate boundaries. Finally, students evaluate the susceptibility of cities around the world to earthquake activity.

Instructional Context

Most earthquakes occur along plate boundaries: release of energy due to relative motion between rocks on adjacent plates causes Earth to shake. Use a model to review the three different plate boundaries. Point out that the area of contact between adjacent plates is where earthquakes are most likely to occur. Displaying a large-scale map of plate boundaries, earthquakes, and volcanoes in the classroom can help students make a geographical connection with the theme of the investigation. For further study, you may want to explore earthquakes that occur along inactive or former plate boundaries.

Where Was That Earthquake?

Library of Congress

KEY CONCEPTS

- Circles representing the distance to an earthquake from three recording stations intersect at the epicenter location.

KEY SKILLS

- Interpreting maps
- Analyzing graphic relationships (intersecting circles)
- Applying information from seismograms to locate an epicenter

ESTIMATED TIME REQUIRED

- 25 minutes Internet use

Investigation Overview

Students examine three seismograms that recorded the 1989 Loma Prieta earthquake, and they review how P and S wave arrival times indicate distance to an epicenter. Students view an animated example showing the process of locating an epicenter: circles with diameters indicated by the seismograms are drawn around each reporting city, and they intersect at the epicenter. Next, students see seismograms from another earthquake event and draw circles of appropriate diameters around each reporting city to locate the epicenter.

Instructional Context

This investigation focuses on how to locate an epicenter once the distances from three recording stations are known. Give students practice calculating distances indicated by P and S wave lag times before they complete this investigation. Draw circles on the board to demonstrate that three recording stations are required for accurate location of an epicenter: show that two intersecting circles have two points in common, but three circles with different center points intersect at only one common point.

Teacher's Guide Chapter 10 Internet Investigation

Which Fault Moved in the Northridge Earthquake?

USGS

Investigation Overview

Students examine a map of faults in the area surrounding Los Angeles, California. They view diagrams depicting fault motion and an animation illustrating movement on a blind thrust fault. Students use epicenter location, ground-shaking maps, ground-motion maps, and location of aftershocks to pinpoint the fault that moved in the 1994 Northridge earthquake.

Instructional Context

In the investigation "Where Was That Earthquake?" (Keycode ES1003), students locate the epicenter of the Northridge earthquake. This investigation offers an opportunity to study the same earthquake further, using detailed data to identify the blind thrust fault that moved in this event. Links at the end of the investigation provide access to photographs showing damage from the Northridge earthquake and other earthquakes in the Los Angeles area.

How Are Mountains Related to Plate Tectonics? ▶ ES1101

NASA

KEY CONCEPTS

- Mountains belts form parallel to current or former convergent plate boundaries.
- Geologic features of active mountain belts can be used to infer plate tectonic settings of ancient mountain belts.

KEY SKILLS

- Interpreting maps and images
- Analyzing geologic features to infer how mountains formed

ESTIMATED TIME REQUIRED

- 35 minutes Internet use

Investigation Overview

In this investigation, students determine the connection between mountain belts and plate boundaries. Students view relief maps and photos of geologic features from the Andes, Himalayan, and Appalachian mountains. Images from the Andes reveal that subduction-related mountain belts are dominated by volcanic and plutonic rocks. The Himalayan images show the predominance of folding associated with continent-continent collision. Students then view images of plutonic rocks and folds from the Appalachian Mountains. From the geologic features, they recognize a similar formation history for the three mountain belts and infer the location of a former convergent plate boundary. Finally, they hypothesize about why mountain belts of similar ages are present on both sides of the Atlantic Ocean.

Instructional Context

Students should be familiar with all three types of plate boundaries, but especially with processes and features that occur at convergent boundaries. Prior to the investigation, review the difference between volcanic and plutonic igneous rocks. Help students recognize that plutonic rocks exposed in ancient mountain belts represent the sub-surface structure of volcanoes that have since been eroded.

How Do Rocks Respond to Stress?

Matthew Nyman, TERC

Investigation Overview

Students view animations showing how rocks respond to stress under differing conditions of temperature and depth. Students sketch the resulting geologic features and indicate the corresponding type of deformation and stress direction. Students then apply their knowledge by analyzing photographs of real geologic features. For each image, they identify the type of deformation, the stress direction that caused it, and the name of the structure.

Instructional Context

Use clay to demonstrate ductile responses to stress: pushing, pulling, and tearing the clay can model the forces of compression, tension, and shear. Break a pencil to model brittle deformation. Ask students to consider what conditions might make the clay respond to force in a brittle manner. Prompt them to list factors that control how an object responds to stress, such as temperature and the rate at which a force is applied. Encourage them to make connections between the clay and rocks.

KEY CONCEPTS

- Stresses associated with plate motion are compression, tension, and shear.
- Brittle deformation refers to rocks breaking.
- Ductile deformation refers to bending or flowing of rocks
- Rocks' response to stress depends on the type of stress applied and conditions of temperature, pressure, and duration.

KEY SKILLS

- Interpreting photographs and diagrams
- Inferring stress direction and conditions from geologic structures

ESTIMATED TIME REQUIRED

- 25 minutes Internet use

What Forces Created These Geologic Features? ▶ ES1106

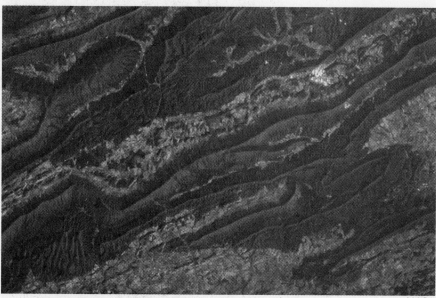

NASA

KEY CONCEPTS

• Structures in mountain belts reveal the direction and types of stresses responsible for mountain building.

KEY SKILLS

• Interpreting images, cross sections, and maps
• Inferring stress directions

ESTIMATED TIME REQUIRED

• 30 minutes Internet use

Investigation Overview:

Students view photographs, maps, and diagrams of geologic features such as folds, thrust faults, and normal faults. They analyze each geologic structure to infer the direction and type of stress (compression, tension, or shear) that created it. Students identify sedimentary layers cut by faults and analyze fault movement to discern the stress that was applied. Students analyze structures from mountain belts in several geologic provinces of the United States.

Instructional Context

Diagrams of faults and folds are often oversimplified. Real geologic structures exposed in roadcuts or depicted as cross sections on geologic maps are more complex and require careful examination to draw conclusions about the forces that created them. Emphasize the importance of correctly identifying the footwall, the hanging wall, and at least one layer on both sides of a fault to infer the force that created it.

Teacher's Guide Chapter 11
Internet Investigation

David H. Harlow, USGS

Teacher's Guide   Unit 3
Internet Investigation

KEY CONCEPTS

- The Earth system responds to volcanic eruptions both locally and globally.
- Volcanic eruptions are the result of ongoing geosphere processes.
- Near an eruption, the biosphere sustains heavy losses.
- The atmosphere acts as a medium to disperse volcanic ash and gases worldwide, decreasing global temperatures.
- The addition of volcanic material to the hydrosphere affects its ability to sustain life.

KEY SKILLS

- Interpreting photographs taken by astronauts
- Predicting trends in interactions among Earth's spheres
- Communicating results

ESTIMATED TIME REQUIRED

- 60 minutes Internet use
- 90 minutes deskwork or homework

Investigation Overview

Students examine images and data to understand how the Earth system responds to an explosive volcanic eruption. Students describe interactions among Earth's spheres for periods before, during, and after the eruption and make predictions about future interactions in the region. A series of before-and-after images illustrating changes to the geosphere, hydrosphere, atmosphere, and biosphere provides students with a concrete point of reference for comprehending the response of each sphere to the eruption. As a culminating project, students develop a presentation to answer the investigation question "How can one volcano change the world?"

Instructional Context

To give students practice (or a review) describing interactions among Earth's spheres, use examples from Chapter 1 Investigations. Rather than focusing on the plate tectonic processes that caused the eruption of Mount Pinatubo, emphasize to students the interconnectedness of all the spheres and their effects on each other. Help students consider how plate tectonic processes that culminate in volcanic eruptions have shaped the geosphere, exerting a primary control on the distribution of Earth's hydrosphere, biosphere, and atmosphere.

How Can One Volcano Change the World?

Unit 3 Project Description

How can one volcano change the world? Students select a volcano to study; they use Web-based resources to gather information and develop a presentation illustrating the effects of eruptions from this volcano on the Earth system. They describe short-term and long-term changes to local and global Earth sphere interactions caused by this volcano. Students present their findings in a format that you have approved.

Criteria for setting student expectations and grading the project are provided below. In conjunction with the criteria, you may also want to use the assessment rubric to evaluate student projects. The rubric is found in the Introduction of the Teacher's Guide.

Unit Project Criteria

1. **A description of the plate tectonic setting of the volcano. (20%)**
   • Provide a map showing plate boundaries, including other volcanoes in the region.
   • Draw a cross-section diagram of the subsurface geology.
   • Describe the volcano's eruption history and magma characteristics.
2. **A description of the Earth system interactions affected by the eruption. (30%)**
   • Identify and describe how the eruption affected the Earth system on a local and global scale.
   • Explain the mechanisms by which Earth sphere interactions are altered by the volcano.
   • Provide supporting images, maps, charts, graphs, or animations.
3. **An explanation of the scientific studies, data, and images used to document changes caused by the volcano. (30%)**
   • Summarize any relevant scientific studies, data, images, or maps which provide evidence of Earth system changes caused by the volcano.
4. **A presentation of research and results. (20%)**
   • Communicate findings and ideas clearly and accurately.
   • Present appropriate data.
   • Use appealing visuals and design.

Alternative Projects

Students may want to choose their own research topic or pursue one of the Questions for Discovery suggested below. The assessment rubric in the Teacher's Guide provides a standard basis for evaluating student performance and is recommended for grading alternative projects.

Sample Questions for Discovery

• How do the effects of explosive volcanic eruptions compare to the effects of non-explosive eruptions?
• Predict how Earth sphere interactions would change if a Mount Pinatubo-sized eruption occurred somewhere on Earth every year.
• What types of preparations could be made to lessen the impacts of volcanic eruptions on humans living near volcanoes?

Teacher's Guide   Unit 3
Internet Investigation

4 UNIT

Earth's Changing Surface

NOAA

Measure the speed of a moving glacier

Compare before-and-after locations of stakes in the surface of a glacier. Calculate the speed of these flowing rivers of ice.

Investigate floods on the Mississippi River

Examine human attempts to control flooding along the Mississippi River. Check how well flood structures worked during major floods over the past century.

Examine storm damage along coastlines

Document the devastation caused by a hurricane on the east coast, and by strong El Niño-related storms on the west coast.

Unit 4
Earth's Changing Surface

Earth Science

4 UNIT Earth's Changing Surface

OTHER WEB RESOURCES

VISUALIZATIONS

Spark new ideas for investigations with images and animations, such as:

• Meandering rivers

• Erupting geysers

• Retreating glaciers

• Seasonal migration of snow line

DATA CENTERS

Extend your investigations with current and archived data and images, such as:

• River maps

• Aquifers

• Glaciers

• Global wind patterns

EARTH SCIENCE NEWS

Relate your investigations to current events around world, such as:

• Flooding

• Coastal erosion

• Active glaciers

• Major dust storms

Unit 4
Earth's Changing Surface

When Is Mud Dangerous? ▶ ES1204

USGS

Investigation Overview

Students examine images of damage caused by mudflows and consider the factors that start mud flowing. They view the results of a muddy experiment on a model hillside and express the relationship between the angle of a hillside and its potential for mudslides. Students suggest other factors that have an impact on whether mud will slide; then they view examples of real mudslides illustrating the conditions they might have predicted. After viewing images depicting other factors affecting mudslides, they write a paragraph to answer the investigation question: When is mud dangerous?

Instructional Context

Many teachers are reluctant to do hands-on experiments with materials as messy as mud. This investigation gives students the opportunity to examine mud and mudflows without getting dirty. The investigation focuses on the natural causes of mudslides. You may want to emphasize how human modifications to hillsides, such as deforestation and excavation for homes and roads, might affect an area's potential for experiencing mudflows. You can augment the investigation by providing images illustrating other types of mass movements (such as landslides and avalanches) and by asking students to compare the conditions that contributed to each result.

How Does Soil Vary from Place to Place?   ▶ ES1206

Arttoday.com

KEY CONCEPTS
- Soil occurs in layers called horizons.
- The depth of topsoil and other horizons varies with location.

KEY SKILLS
- Measuring topsoil depths
- Plotting topsoil depth versus average annual precipitation
- Comparing features of soils

ESTIMATED TIME REQUIRED
- 35 minutes Internet use

Investigation Overview

In this investigation, students analyze soil profiles to determine how soil varies from one place to another. For one soil, students draw lines to indicate boundaries between soil horizons, or layers, and they describe the soil in each horizon. Then, students measure the topsoil depth for six state soils. Students plot topsoil depth versus average annual precipitation to look for a possible relationship. Finally, they access a database of state soil profiles, estimate the depth of their own state topsoil, and compare its location on the depth-versus-precipitation graph with the other sample soils.

Instructional Context

Soil is a naturally occurring, unconsolidated assemblage of organic material and weathered rock fragments. The soils considered in this investigation show a rough trend of topsoil depth increasing with an increase in average annual precipitation. You may want to have students investigate other soils to see if the trend holds true, or have them consider other factors that may affect the depth of topsoil. If possible, arrange to have students handle actual soil samples in addition to viewing the images of soil. Use a microscope to examine the constituents of your local soil, or dig into an undisturbed area to expose a soil profile.

Teacher's Guide   Chapter 12
Internet Investigation

How Does Stream Flow Change over Time? ▶ **ES1301**

Andrea Booher, FEMA

KEY CONCEPTS

- Stream discharge is the volume of water that flows past a point over a specific period of time.
- Factors that affect stream discharge include rainfall, snowmelt, and releasing water from dams.

KEY SKILLS

- Interpreting graphs
- Analyzing timing of events from graphs
- Inferring causes of changes in stream discharge

ESTIMATED TIME REQUIRED

- 40 minutes Internet use

Investigation Overview

Students view images that suggest the key controls of stream flow. They view examples illustrating situations where rainfall, snowmelt, and water release from dams caused stream flow to change. Students examine graphs of stream discharge coupled with graphs of rainfall, snow depth, and temperature to analyze how these factors affect stream flow. Finally, students interpret a set of graphs to make inferences about the factors controlling observed discharge.

Instructional Context

You can introduce the concept of measuring stream discharge by using water flow from a faucet. Have students calculate faucet discharge by timing how long it takes to fill a one-gallon container. The final question of the investigation asks students to infer the cause of changes in stream discharge for a full month. Encourage them to pay close attention to the graphs and write a complete summary citing specific evidence to support their conclusions.

What Controls the Shape of a Delta?

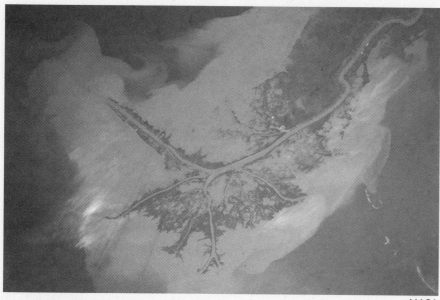

NASA

KEY CONCEPTS

- The size and shape of river deltas are controlled by river discharge, amount of sediment, and energy associated with waves and tides in the basin.
- Rivers, waves, and tides interact to shape deltas. One of these elements is usually dominant over the others.

KEY SKILLS

- Interpreting images
- Inferring dominant processes

ESTIMATED TIME REQUIRED

- 30 minutes Internet use

Investigation Overview

Students examine images of deltas from around the world and analyze the factors controlling delta shape. The investigation focuses on how delta shape reflects the environment where the sediment is deposited. Students analyze photographs of deltas and categorize the delta-shaping processes as river-dominated, wave-dominated, or tide-dominated.

Instructional Context

Students should be aware of the basic concept that rivers transport sediment. Deltas form where sediment is deposited in a low-energy body of water such as a lake or ocean basin. How the sediments accumulate depends on how much sediment is deposited and how tides and waves in the body of water interact with the sediments. You can model delta formation on a stream table or with a hose in an area of loose sediments, or watch for small-scale deltas that form in puddles after rain.

Teacher's Guide Chapter 13
Internet Investigation

Have Flood Controls on the Mississippi River Been Successful?

FEMA

KEY CONCEPTS

- Humans have attempted to manage flooding rivers for centuries, with varying degrees of success.

KEY SKILLS

- Interpreting images
- Inferring how flood management structures work

ESTIMATED TIME REQUIRED

- 45 minutes Internet use

Investigation Overview

Students examine the Mississippi River as a case study to illustrate human attempts at managing flooding rivers. Students observe structures employed in the effort to control river flow, and follow the chronology of flood management along the Mississippi River, including particulars of three significant flooding events. Students examine images and details of the events and evaluate the effectiveness of flood management. Finally, students are introduced to recent techniques for river management.

Instructional Context

A long history of flood management efforts is documented for the Mississippi River. Excellent Websites offer insight into flooding events, flood management control plans, and historic figures involved with the cause. Encourage discussion of the causes of flooding and flood hazards. Watch for current flooding events covered in the news, and analyze the river system and any attempts that have been made to control it.

How Does Water Move Through the Ground?   ▶ ES1401

Jennifer Loomis, TERC

KEY CONCEPTS

- Groundwater moves between inter-connected pore spaces or cracks in rocks.
- Sandstone and limestone have higher permeability than shale.

KEY SKILLS

- Visualizing how water moves through the ground
- Analyzing flow rates of groundwater through common rock layers

ESTIMATED TIME REQUIRED

- 35 minutes Internet use

Investigation Overview

Students use an interactive animation to visualize water contained within and flowing through different rock materials. Students compare the flow rates of groundwater pumped from sandstone, limestone, and shale and then describe porosity and permeability for each rock type. Finally, students describe characteristics that make each type of rock a desirable or undesirable aquifer for supplying human water needs.

Instructional Context

Though relatively few students have familiarity with wells, dramatic increases in the populations of arid regions means that increasing numbers of humans depend on groundwater for their survival. Availability of uncontaminated water at accessible depths is an issue that will affect many towns and cities. Find out what types of aquifers are below ground where you live—you may be able to check the flow rates of private or municipal wells in your region.

Teacher's Guide Chapter 14
Internet Investigation

How Many People Can an Aquifer Support? ▶ ES1406

Denver Metro Convention and Visitors Bureau

KEY CONCEPTS

- Aquifers are recharged by precipitation and discharged by pumping for human use.
- Changes in the water level of an aquifer reflect the relative amounts of recharge and discharge.

KEY SKILLS

- Analyzing factors controlling recharge and discharge of an aquifer
- Predicting changes in water use
- Hypothesizing about the cause of water level changes

ESTIMATED TIME REQUIRED

- 30 minutes Internet use
- 10 minutes deskwork or homework

Investigation Overview

Students investigate the water level of the Denver Basin aquifer system. They begin by focusing on a decline in the water level of the Arapahoe aquifer (part of the Denver Basin aquifer system) between 1890 and 1960. To understand the system, students consider factors that control an aquifer's water level. They compare water levels in the aquifer over time with precipitation, population, and water-use data. Students predict how water use patterns might change as land formerly used to grow irrigated crops is covered by homes and businesses. Students examine data showing that the water level of the Arapahoe aquifer has increased since 1960, and they hypothesize about what caused the trend to reverse. Finally, they calculate a rough estimate of the number of people the aquifer might support.

Instructional Context

The public water supply for many homes and municipalities across the country comes from groundwater. Find out if groundwater is used in your area, and how it is pumped out of the ground. Visit or show photographs of a well in your region to raise student awareness of their dependence on groundwater. Ask students to consider how long it takes for water to accumulate in aquifers; emphasize that available groundwater is a limited resource. Though aquifers are complex systems, students can compare recharge and discharge rates to draw basic conclusions.

Teacher's Guide Chapter 14
Internet Investigation

Copyright © McDougal Littell Inc.

How Does Land Cover Affect Global Temperature?

Tom Kellogg, onboard the US Coast Guard ice breaker *Glacier*

KEY CONCEPTS

• The amount of sunlight reflected from Earth's surface (albedo) varies with land cover.
• Increases in snow and ice cover can intensify global cooling.
• Decreases in snow and ice cover can intensify global warming.

KEY SKILLS

• Ranking the albedo of common types of land cover
• Predicting how changes in land cover might affect global temperatures

ESTIMATED TIME REQUIRED

• 35 minutes Internet use

Investigation Overview

In this investigation, students predict how changing land cover might affect the planet's temperature balance. Students are introduced to the concept of albedo by ranking the amount of sunlight reflected from four common types of land cover. They discover that a change in the amount of reflected energy alters the amount of absorbed energy, ultimately affecting global temperature. Students conclude that changes in areas covered by snow or ice can intensify warming or cooling periods.

Instructional Context

This investigation concentrates on the role of land cover in determining the albedo of a location—the effects of clouds or aerosols in the atmosphere are not examined here. A simple demonstration can help illustrate how areas with higher reflectivity absorb less heat: Attach identical thermometers to two ring stands, and place them at the same distance from a light bulb. Put a sheet of white construction paper in front of one thermometer and a piece of black construction paper of similar weight in front of the other. Turn on the light and record temperatures from each thermometer for several minutes; then discuss your results. Ask students to explain conditions in nature the model represents.

Teacher's Guide Chapter 15
Internet Investigation

How Fast Do Glaciers Flow?

Peggy O'Neill, NASA, GSFC

Investigation Overview

Students consider how to measure the speed of a moving glacier; then they use data gathered by scientists to do so. Students analyze movement of survey flags on the surface of Mathes Glacier in Alaska to calculate its rate of flow. They analyze vector data to visualize the difference in flow rates across the glacier. Students also examine an image depicting flow rates for Lambert Glacier in Antarctica. Finally, they compare the rate of movement of the two glaciers.

Instructional Context

Though ice is a solid, glaciers flow under the force of gravity. The speed at which a glacier flows depends in part on the land surface over which it is flowing. Overall, most glaciers move quite slowly—only centimeters per day. Flow rates vary across the width and along the length of individual glaciers. Research programs around the world regularly monitor and analyze the movement of glaciers.

What Controls the Shape and Motion of Sand Dunes?

Matthew Nyman, TERC

KEY CONCEPTS

- Dune formation requires loose sediments, energy to move them, obstacles to act as a windbreak, and a dry climate.
- The types of dunes that form depend on sediment supply and wind direction.

KEY SKILLS

- Identifying conditions required for dune formation
- Interpreting images
- Inferring wind direction and sediment supply for dunes

ESTIMATED TIME REQUIRED

- 45 minutes Internet use

Investigation Overview

By observing dune images and animations, students investigate conditions responsible for the formation of different kinds of dunes. They infer wind direction and sediment supply from images of dunes. Students document the presence of dunes on Mars and the absence of dunes on other planets and moons. They speculate about environmental conditions on other planets.

Instructional Context

Dunes form where loose sediments, sand-sized or smaller, accumulate around a wind break and begin to pile up behind it. As sand accumulates, the size of the wind break increases. If there is a supply of sediments, the dunes grow. Dunes and dune fields are dynamic landscapes that change rapidly. Help students discern between the shallow slope on the windward side of dunes and the steeper slope on the leeward side.

**Teacher's Guide   Chapter 16
Internet Investigation**

Where Did This Sand Come From?

NOAA

KEY CONCEPTS

• Sand composition varies from place to place.
• Grains within sand samples can indicate the source rock of the sand.

KEY SKILLS

• Interpreting images
• Inferring sources of sand grains

ESTIMATED TIME REQUIRED

• 30 minutes Internet use

Investigation Overview

Students view images of sand grains from different sources, focusing on specific characteristics that indicate their origins. Students study samples, relating shell and coral fragments to a nearby marine source and volcanic glass or vesicular grains to a volcanic source. Students examine eight samples of sand of unknown origin. They analyze grains within each sample to determine the source areas.

Instructional Context

Sand found on beaches, along riverbanks, and in deserts often has a simple composition that reflects where it formed. Shell and coral fragments indicate a nearby marine source. Volcanic glass fragments indicate a nearby volcanic source. Sand that has been sorted and transported long distances contains only the most durable minerals, such as quartz, feldspar, and magnetite. Collecting sand from locations around the world is a hobby for some people. The sand images used here were provided by the International Sand Collectors Society.

How Do Storms Affect Coastlines?

NOAA

KEY CONCEPTS

- Interactions among the hydrosphere, geosphere, and atmosphere affect coastlines.
- Storm events have a major impact on coastal areas.

KEY SKILLS

- Interpreting images
- Describing change over time

ESTIMATED TIME REQUIRED

- 45 minutes Internet use

Investigation Overview

Students examine sets of before-and-after images to investigate how storms affect coastal areas. Images from areas of North Carolina that were in the path of Hurricane Dennis in 1999 are used to document the affect of hurricanes. Images of coastal areas in California are used to investigate changes caused by intense winter storms from the 1997–98 El Niño event.

Instructional Context

A majority of the world's population lives in coastal regions, so these areas are the subjects of intense study. Coasts are dynamic environments subject to gradual as well as rapid change. Tectonic forces and normal erosion and deposition patterns change coasts gradually; storms can change a coastline drastically in a matter of hours or days. Discuss some of the weather conditions associated with hurricanes on the east coast and with El Niño-related weather phenomena on the west coast to give students atmospheric context for these events that affect coasts.

What are the Costs and Benefits of Damming a River?

ArtToday.com

Investigation Overview

In this investigation, students study the impacts resulting from the construction of Glen Canyon Dam on the Colorado River in Arizona. Students also explore Lake Powell, the body of water formed by the damming of the river. Students evaluate the influence of the dam and lake on six different aspects of life in the region, then rate each as a cost or a benefit. Taken together, the ratings represent a quantifed evaluation of the dam's impact on the area. Students consider how well the quantified evaluation matches their overall assessment of the dam. They consider modifications to the rating system, such as adding new aspects for consideration or assigning weighting factors to aspects they deem more or less important than others. As a culminating project, students select another dam and investigate how the Earth system has changed as a result of building it.

Instructional Context

In 1957, when Glen Canyon Dam was constructed, the long-term consequences of changing the Earth system by building large dams were not well understood. Today, several groups actively protest the existence of Glen Canyon Dam and Lake Powell, arguing that nature should regain control of this area that humans have changed so drastically. Positive effects of the dam include electric power, recreation, water control, and a strong economy in the area. The negative impacts of the dam include the loss of historical cultural sites, dramatic changes in the flora and fauna upstream and downstream from the dam, and a lost canyon wilderness now buried in lake sediments.

What are the Costs and Benefits of Damming a River?

Unit 4 Project Description

How are Earth sphere interactions altered by the damming of rivers? Using Web resources, students identify a major river that has been dammed. They obtain information and scientific studies on the dam and how it has affected the river system and changed interactions among Earth's spheres. Students present their findings in a format that you have approved.

Criteria for setting student expectations and grading the project are provided below. In conjunction with the criteria, you may also want to use the assessment rubric to evaluate student projects. The rubric is found in the Introduction of the Teacher's Guide.

Unit Project Criteria

1. **A description of the river system and the dam. (20%)**
   - Describe the entire river system. Include a general description of the landscape through which the river flows, the major sources of its water, annual flow patterns, plant and animal ecology, and human uses.
   - Provide a description of the dam and information about why the dam was built.
   - Develop a descriptive map with representative images of the river system.
2. **A description of changes to the river system resulting from the dam. (30%)**
   - Identify and describe changes to the river system caused by the dam.
   - Provide supporting images, maps, charts, or graphs to illustrate these changes.
3. **A summary of scientific studies on the changes caused by the dam. (20%)**
   - Summarize studies and information about the changes caused by the dam.
   - Summarize pertinent studies from different rivers and explain how the findings from these studies might be applied to this river.
4. **An evaluation of the changes caused by the dam. (10%)**
   - Evaluate each of the identified changes as positive or negative.
   - Present appropriate information that supports these ratings.
5. **A presentation of research findings. (20%)**
   - Communicate findings and ideas clearly and accurately.
   - Use appealing visuals and design.

Alternative Projects

Students may want to choose their own research topic or pursue one of the Questions for Discovery suggested below. The assessment rubric in the Teacher's Guide provides a standard basis for evaluating student performance and is recommended for grading alternative projects.

Sample Questions for Discovery

- What are the short-term and long-term effects on the Earth system of removing existing dams from rivers?
- How have the deltas and floodplains of major world rivers been affected by dams?
- How might some of the negative impacts on the ecosystems of dammed rivers be decreased?

5 UNIT Atmosphere and Weather

NASA

Track a hurricane

Get the latest satellite images of these massive storms and track their movement.

Make your own weather forecast

Use current maps of cloud cover and precipitation for your local area to forecast tomorrow's weather.

Investigate global climate change

Read the record of climate change recorded in Antarctica's ice. Compare today's changing climate with climate changes over the last 400,000 years.

Unit 5
Atmosphere and Weather

5 UNIT Atmosphere and Weather

OTHER WEB RESOURCES

VISUALIZATIONS

Spark new ideas for investigations with images and animations, such as:

• Clouds forming and dissipating

• Weather systems

• Forest fires

• Auroras

DATA CENTERS

Extend your investigations with current and archived data and images, such as:

• Satellite images of cloud cover

• Radar images of precipitation

• Global and regional temperature maps

• Safety tips for hurricanes, tornadoes and snow storms

EARTH SCIENCE NEWS

Relate your investigations to current events around world, such as:

• Hurricanes

• Tornadoes

• Extremely hot or cold weather

• Forest fires

**Unit 5
Atmosphere and Weather**

What Can You Learn from a Thermometer on a Rising Balloon?

Emmett Given, NASA/MSFC

KEY CONCEPTS

- Instruments carried by high altitude balloons record data that reveal the layered structure of Earth's atmosphere.
- The atmosphere is divided into several discrete layers, based on measurable atmospheric phenomena.

KEY SKILLS

- Interpreting graphs
- Analyzing physical attributes of the atmosphere in a vertical profile.

ESTIMATED TIME REQUIRED

- 30 minutes Internet use

Investigation Overview

Students explore temperature data generated by instrument packages attached to high altitude balloons. They look at patterns of changing temperature to infer the existence of distinct boundaries between layers in the atmosphere. To investigate the global shape of atmospheric layers, students compare temperature data with altitude data collected at different latitudes and at different times of day. Students also consider how wind speed and direction change at altitudes they identify as layer boundaries.

Instructional Context

High altitude balloons carry a variety of scientific instruments into the atmosphere, recording data at regular time intervals as they rise. Government agencies and groups of amateur weather watchers launch these balloons to monitor conditions in the atmosphere's upper levels. Most commonly, instruments attached to these balloons transmit data that are used to create a profile of temperature as it changes with altitude. Balloon-based instruments also measure wind speed, wind direction, and radiation. Some even carry cameras that capture both visible and infrared images. These instrument packages eventually fall back to Earth—if possible, show such an instrument package in class, or discuss the possibility of finding one based on an average of 1000 balloon launches each day in the United States.

How Does the Temperature at One Location Change over a Year?

Bryan Aivazian, TERC

KEY CONCEPTS

- Locations on Earth receive predictable amounts of incoming solar radiation throughout the year.
- Insolation, elevation, and proximity to a coastline are factors that control surface temperature.

KEY SKILLS

- Interpreting satellite animations
- Quantifying observations through measurement
- Analyzing environmental attributes over time and space
- Predicting temperature based on elevation

ESTIMATED TIME REQUIRED

- 35 minutes Internet use

Investigation Overview

Students explore animations of insolation (incoming solar radiation) and global surface temperature over a year. They discover that temperature is strongly influenced by insolation, resulting in seasonal temperature changes, yet other factors affect the observed temperature as well. Students quantify the effects of elevation on temperature for locations at similar latitudes. They also examine temperatures for locations at similar latitudes and elevations differing only in the locations' proximity to a coastline.

Instructional Context

If incoming solar radiation were the sole factor controlling surface temperature, all locations at the same latitude would have the same temperature. The complexity of observed surface temperatures provides evidence that several factors besides insolation are at work. After investigating factors that control temperature, you may want to analyze daily temperatures, weather conditions, and hours of daylight for your location over a year. These data would reveal factors that control seasonal temperature changes at a single location.

How Does the Ozone Layer Change over Time? ▶ ES1708

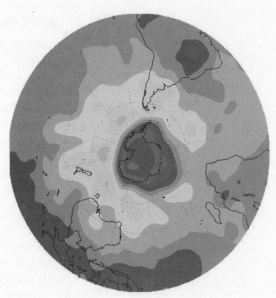

Climate Prediction Center/NCEP/NWS/NOAA

Investigation Overview

Students animate a series of global images showing monthly atmospheric ozone levels. They examine patterns of annual variation for the northern and southern polar regions. Another series of images shows the average ozone level measured each October from 1979 through 1992. Students compare these images to see how ozone levels have changed over time. Students measure the major and minor axes of the Antarctic ozone hole in these images to quantify changes in its size. They also record the minimum levels of ozone in the atmosphere for a 14-year period. Finally, students examine recent trends in ozone levels, looking for changes following the establishment of the Montreal Protocol, an international agreement to reduce production of chlorofluorocarbons (CFCs).

Instructional Context

Global ozone concentrations are measured by satellite instruments from Earth orbit. Ozone is measured by comparing the atmosphere's ability to absorb and reflect specific wavelengths of solar radiation. Each year, levels of ozone over the Southern Hemisphere drop to a minimum during October. Analyzing images from successive Octobers allows students to quantify how minimum levels have changed over the past two decades.

Which Way Does the Wind Blow?

NOAA

KEY CONCEPTS

- Wind direction can be deduced from vegetation and rainfall patterns.
- The windward sides of mountains receive more rain than the leeward sides.

KEY SKILLS

- Interpreting images
- Analyzing rainfall patterns
- Predicting wind direction

ESTIMATED TIME REQUIRED

- 30 minutes Internet use

Investigation Overview

Students examine vegetation patterns, annual rainfall amounts, and topography of the island of Oahu to infer the general direction of wind blowing over the island. They view an animation that demonstrates orographic (mountain-related) uplift of air to discover that the windward sides of mountains receive more rain than the leeward sides. With this knowledge, students analyze annual rainfall amounts on Oahu to differentiate the windward from the leeward side of the island. Finally, students examine rainfall patterns and topography of California to predict the general wind direction in another setting.

Instructional Context

Vegetation patterns revealed in photographs taken by astronauts are evidence of interactions among Earth's spheres. The atmosphere applies hydrosphere to the geosphere on the windward side of mountains, nurturing the biosphere. Challenge students to recognize, in images from space, the results of atmospheric processes revealed by the other spheres. The collection of space shuttle images is extensive, engaging, and suitable for a wide variety of aerial pattern studies. Visit NASA's main Website to locate and download photographs taken by astronauts.

Teacher's Guide Chapter 18
Internet Investigation

How Acidic Is Your Rain?

Freefoto.com

Investigation Overview

Students examine images showing the chemical nature of rain that falls on the contiguous United States. The images depict pH (acidity or alkalinity) of rainwater as well as concentrations of sulfate, nitrate, and ammonium ions. Students view animations of these images to study changes in rain chemistry over time. They also correlate observed chemical patterns with the locations of existing coal-fired power plants. To investigate the acidity of rain in their own region, students retrieve precipitation data from the National Atmospheric Deposition Program's Web site. Note that the external Web site requires students to log in before they can retrieve data. Students do not provide any information that could be used to identify them—they only indicate that the data will be used in a school setting. Students graph the data to explore trends in their local precipitation chemistry.

Instructional Context

Chemical data from a widespread network of collection and recording sites allow researchers to monitor changes in precipitation chemistry on a variety of time scales (annual, seasonal, monthly, weekly, and daily). These data are freely available to the public. Familiarize your students with the pH scale before they begin the investigation. You may want to collect samples of water from various sources such as rain, lakes, bottled water, and tap water and test them with a pH meter.

How Does the Jet Stream Change through the Year?

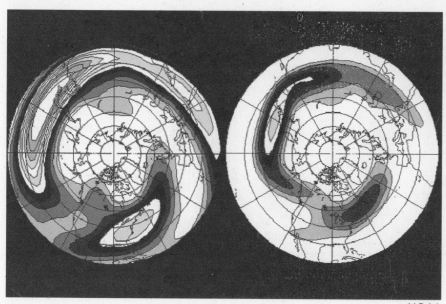

NOAA

KEY CONCEPTS

- Jet streams are areas of strong winds in the upper atmosphere.
- The location and intensity of jet streams vary seasonally.
- Weather patterns are influenced by jet streams.

KEY SKILLS

- Interpreting maps
- Analyzing change over time
- Predicting southern hemisphere patterns based on northern hemisphere patterns

ESTIMATED TIME REQUIRED

- 30 minutes Internet use

Investigation Overview

Students examine an animation of jet stream maps to discover changes in the general location and intensity of the northern hemisphere polar jet stream over the course of a year. Students learn why the jet stream changes and how it can affect surface weather conditions. They explore a jet stream analysis site that shows the current location of the jet stream and associated weather conditions. Students synthesize their observations by sketching the average summer and winter paths of the jet stream across North America.

Instructional Context

This investigation focuses on the northern hemisphere polar jet stream. Other jet streams include a southern polar jet stream as well as weaker, low-latitude tropical jet streams in both hemispheres. The northern hemisphere polar jet stream follows a fairly predictable cycle over the course of a year. Strong temperature differences between cold polar air and warmer tropical air cause a more intense jet stream during the winter months than in summer. Because of its strong effect on surface weather conditions, the location of the polar jet stream can often be detected on surface weather maps.

Teacher's Guide Chapter 19
Internet Investigation

Could You Break the Record for an Around-the-World Balloon Flight?

▶ **ES1908**

Sukree Sukplang, Reuters 1999

Investigation Overview

Students explore information about Breitling Orbiter 3's successful around-the-world balloon flight and other notable, but unsuccessful, attempts. They discover how balloons are steered—by rising or descending into desirable air currents to control their direction and speed. Students then explore the relationship between jet streams and the flight paths of balloons. To conclude, students examine current jet stream maps and determine if these jet streams offer the conditions that would be necessary to beat the record time for an around-the-world balloon flight.

Instructional Context

Even though balloons were the first form of human flight, it was not until 1999 that a balloon was able to complete a non-stop flight around the world. Technologically advanced equipment and an ability to precisely predict the path of jet streams were both critical to this achievement. Meteorologists at the balloon team's mission control station constantly analyzed the position and intensity of atmospheric winds and weather patterns. They directed the pilots to steer up or down into jet streams that would move the balloon toward its goal.

Many of the same jet stream analysis tools used by balloon teams are available on the Internet. Armed with up-to-date information, students will be able to predict whether current jet stream conditions are conducive to the completion of an around-the-world balloon flight attempt.

How Does a Mid-Latitude Low Develop into a Storm System?

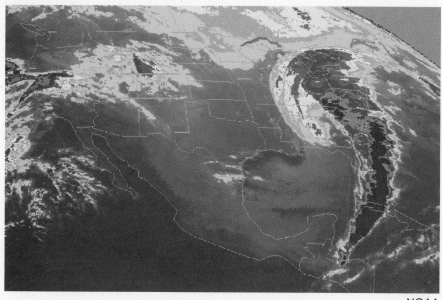

NOAA

KEY CONCEPTS

- Low-pressure systems result from the interaction of warm and cold air masses.
- Storms and precipitation occur along the boundary of interacting warm and cold air masses.
- Satellite imagery is used to help predict weather.

KEY SKILLS

- Interpreting color-enhanced satellite imagery
- Comparing surface weather data with corresponding satellite imagery
- Predicting storm progression

ESTIMATED TIME REQUIRED

- 45 minutes Internet use
- 30 minutes deskwork or homework

Investigation Overview

A low-pressure storm system so powerful that it came to be known as the "Storm of the Century" rocked the eastern United States in March 1993. Students begin by examining generalized processes and conditions that lead to the development of mid-latitude storm systems. They then track the life cycle of this classic mid-latitude storm from its early formation, through its most intense period, to its dissipation as it moved from the Gulf of Mexico up the eastern seaboard. By comparing surface weather observations with a series of corresponding color-enhanced satellite images, students discover the characteristics of mid-latitude storm systems.

Instructional Context

Mid-latitude low-pressure systems are the most common weather system in the U.S. The March 1993 storm developed when a strong polar jet stream pushed a cold air mass into a northerly moving warm air mass over the Gulf of Mexico. The result was a case-study example of an intense mid-latitude storm system. A sequence of color-enhanced infrared satellite images clearly shows the life cycle of this storm. By closely comparing surface weather observations from several cities with corresponding satellite images, students discover how weather conditions change as the storm approaches, passes through, and moves beyond a given location.

Teacher's Guide Chapter 20
Internet Investigation

NASA

Investigation Overview

Students apply their understanding of weather processes to develop a prediction for the next day's weather. First, they discover the wide variety of weather data available by exploring different kinds of satellite imagery, surface weather maps, and weather forecast maps. Using a selected set of links to various weather data sources, students gather information that will help them predict their regional weather for the next day. On the following day, they collect weather data and critically review the accuracy of their predictions.

Instructional Context

A vast amount of current weather data is available for public use on the Internet. While these data can be informative to the general public, they convey much more meaning when coupled with some simple instruction on the interpretation and use of these resources. Help students become familiar with the standardized weather symbols and general methods for interpreting satellite images. Encourage them to make connections between what they see on a weather map and the conditions they observe outside.

What Factors Control Your Local Climate?   ▶ ES2101

NPS

KEY CONCEPTS

- Climate is the long-term average of weather conditions for a given location.
- Some climate control factors influence local climate more than others.

KEY SKILLS

- Interpreting graphs of climate information
- Comparing climate data from different locations
- Predicting factors that control local climate

ESTIMATED TIME REQUIRED

- 45 minutes Internet use
- 20 minutes deskwork or homework

Investigation Overview

Students gain an understanding about the factors that control local climate. They distinguish the relative degree of influence that various climate control factors have on local climates by examining climate data for several pairs of cities. By comparing the climates of different locations, students see that some climate control factors exert greater influence on local climate than others. Students then develop a more complete knowledge of their local climate by reviewing published climate data for their own area. They examine maps and other sources of climate information to determine which factors have the most influence on their local climate.

Instructional Context

The factors that control local climate include latitude, elevation, nearby water, ocean currents, topography, prevailing winds, and vegetation. Elevation and latitude are generally the most obvious factors influencing local climate. Large bodies of nearby water, particularly oceans, also play major roles in controlling local climate. The degree of influence that topography, prevailing winds, and vegetation have on local climate is generally more subtle than other factors. The local climates of several paired groups of cities are compared as a way to highlight how specific control factors influence climate. By looking at the most variable climate control factor(s) between each of the paired cities, students will be able to see which climate control factors are most influential.

Teacher's Guide Chapter 21
Internet Investigation

How Do Ice Cores of Glaciers Tell Us about Past Climates?

NOAA

Copyright © McDougal Littell Inc.

KEY CONCEPTS

- Ice cores of glaciers provide indirect, but detailed, data about past climatic conditions.
- Climatic conditions have varied over the past 190,000 years.
- Scientists can use knowledge about past climate to help predict future climate changes.

KEY SKILLS

- Interpreting graphs
- Inferring past climate trends.
- Predicting future climate trends

ESTIMATED TIME REQUIRED

- 30 minutes Internet use
- 20 minutes deskwork or homework

Investigation Overview

Students learn how ice cores from glaciers provide detailed information about past climate conditions. Students interpret and annotate graphs showing how temperature, carbon dioxide (CO_2), and dust concentrations have varied over the past 190,000 years. By comparing their annotated graphs, they discover the relationships among temperature fluctuations, greenhouse gas concentrations, and atmospheric dust. Finally, students are asked to make inferences about past climates and predictions about how climate will change in the future.

Instructional Context

Ice cores from glaciers supply one of the most detailed pictures of Earth's climate over the past half million years. Data derived from the Vostok, Antarctica deep ice core show that changes in atmospheric CO_2 concentration parallel variations in temperature. CO_2 is considered a greenhouse gas because it plays a role in trapping sunlight energy within Earth's atmosphere. The data also show that an increase in CO_2 levels frequently precedes increases in temperature, possibly indicating a causal relationship between these two variables. Scientists can use information gathered from ice cores to try to predict the effects that recent changes in atmospheric greenhouse gas may have on global climate.

Teacher's Guide Chapter 21
Internet Investigation

How Might Global Climate Change Affect Life on Earth?

US Army Corps of Engineers

KEY CONCEPTS

- Scientific observations and analyses indicate that human activities are contributing to global climate change.
- Scientists develop models that help them make predictions about future changes.
- Impacts of global climate change will likely affect the entire Earth system.

KEY SKILLS

- Interpreting graphs and charts
- Synthesizing information from a variety of sources
- Evaluating the implications of climate change models
- Explaining how an expected impact of global climate change will affect the Earth system.

ESTIMATED TIME REQUIRED

- 90 minutes Internet use
- 90 minutes deskwork or homework

Investigation Overview

Students begin by examining scientific evidence of global warming and considering the role that humans play in influencing climate. Students analyze graphs generated by scientific models for predicting future climatic conditions. Using these models, students describe possible impacts of a changing climate. Students delve more deeply into these issues as they explore a series of Internet sites focusing on world climate change and its consequences. To conclude, students are asked to suggest possible solutions that might forestall or decrease the projected impacts of global climate change.

Instructional Context

Scientific research and analysis have largely settled the debate about whether or not Earth's climate is warming; global warming is now recognized as a reality. Scientific (and political) debate now centers on the degree to which human activities are responsible for these observed climate changes, as well as what can and should be done about it. Scientific models that predict changes to Earth systems are largely based upon different scenarios for the amount of greenhouse gases present in the atmosphere. The implications surrounding many projected impacts are profound and could significantly influence the world in the next several decades.

Teacher's Guide Unit 5
Internet Investigation

How Might Global Climate Change Affect Life on Earth?

Unit 5 Project Description

How will processes in the Earth system change in response to global climate change? Using Web resources, students identify an impact that is expected to occur as a result of global climate change. Examples of impacts include a rise in sea level, increased frequency of severe weather events, and changes in biodiversity. Students develop a presentation that explains the impact and how it will change the Earth system. Students present their findings in a format that you have approved.

Criteria for setting student expectations and grading the project are provided below. In conjunction with the criteria, you may also want to use the assessment rubric to evaluate student projects. The rubric is found in the Introduction of the Teacher's Guide.

Unit Project Criteria

1. **A description of the impact caused by global climate change. (20%)**
   - Explain the nature of the impact and its expected location, timing, and duration.
   - Use images, graphics, or animations to illustrate the impact.
2. **A description showing how global climate change affects the Earth system. (30%)**
   - Identify portions of the Earth system affected by climate change.
   - Describe how climate change alters the normal functioning of the system and leads to the impact.
   - Use images, graphics, or animations to illustrate the changes in Earth system function.
3. **An explanation of the scientific evidence linking the impact to global climate change. (20%)**
   - Describe the scientific evidence for the relationship between the impact and global climate change.
   - Identify strengths and any uncertainties in the scientific evidence.
4. **A recommendation for decreasing the degree of the impact. (10%)**
   - Explain ways that the impact could be averted or decreased.
5. **A presentation of research and results. (20%)**
   - Communicate findings and ideas clearly and accurately.
   - Present appropriate data.
   - Use appealing visuals and design.

Alternative Projects

Students may want to choose their own research topic or pursue one of the Questions for Discovery suggested below. The assessment rubric in the Teacher's Guide provides a standard basis for evaluating student performance and is recommended for grading alternative projects.

Sample Questions for Discovery

- How might the Earth system react to increasing amounts of atmospheric carbon dioxide?
- Which atmospheric components and processes increase global warming, and which components and processes work to cool the atmosphere?
- What scientific evidence points to human activities as a major cause of global climate change, and what evidence indicates that global climate change is a natural process?

6 UNIT Earth's Oceans

NOAA

Drain the ocean to explore its floor

Use visualization tools to drain the oceans, revealing mountain chains and deep trenches.

Chart a path to sail around the world

Use your knowledge of ocean currents and wind patterns to plot the fastest path for sailing around the world.

Find the best fishing spots in the Atlantic

Analyze sea surface temperature images of the Gulf Stream in the Atlantic Ocean to choose the best locations for fishing.

Unit 6
Earth's Oceans

6 UNIT Earth's Oceans

OTHER WEB RESOURCES

VISUALIZATIONS

Spark new ideas for investigations with images and animations, such as:

• Global patterns of currents

• Upwelling

• Black smokers

• Islands and atolls

DATA CENTERS

Extend your investigations with current and archived data and images, such as:

• Ocean currents

• Ocean floor

• Coral reefs

• Tides

EARTH SCIENCE NEWS

Relate your investigations to current events around world, such as:

• El Niño

• Sea level rise

• Undersea exploration

• Monsoons

Unit 6
Earth's Oceans

Copyright © McDougal Littell Inc.

How Do Temperature and Salinity Affect Mixing in the Ocean?

NOAA

Investigation Overview

Students observe maps showing average annual sea temperatures and salinity concentrations to identify relationships and anomalies. They discover that warm, isolated ocean waters have higher salinity and cold, well-mixed ocean waters have lower salinity. Using maps and a temperature-salinity graph, students determine water density for various ocean locations and predict what will happen when these waters mix.

Instructional Context

This investigation gives students the opportunity to apply the concepts of salinity and density presented in Chapter 22. A simple classroom demonstration—pouring a colored saltwater solution into a beaker of clear water—can set the stage for this investigation. Students can apply what they've seen in the beaker to the world's oceans. Point out that mixing of ocean waters is essential to the support of marine ecosystems. Phytoplankton are generally concentrated in cold waters. Mixing of ocean waters helps distribute this resource that serves as the base of the food chain in oceans.

KEY CONCEPTS

- Salinity is generally higher in areas of warmer waters and in isolated bodies of water.
- Salinity is generally lower in areas of colder waters and along continents.
- Denser water sinks when it mixes with less dense water.
- Less dense water rises when it mixes with denser water.

KEY SKILLS

- Identifying relationships between sea temperature and salinity
- Interpreting seawater density from a graph
- Predicting how water will mix

ESTIMATED TIME REQUIRED

- 35 minutes Internet use.

Teacher's Guide Chapter 22
Internet Investigation

What's Responsible for Smaller Shrimp Catches? ▶ ES2206

NOAA

Investigation Overview

Students observe maps of mean oxygen concentration in the Gulf of Mexico for each season of the year. They observe decreases in oxygen concentration during the summer and learn about dead zones, areas of the oceans that have been rendered virtually lifeless because of severe oxygen depletion (hypoxia). Students learn that some scientists believe this phenomenon is caused by the high levels of nitrates, originating in the Mississippi River Basin farming areas as fertilizers and organic forms of nitrogen, washing into the Gulf. Students observe maps showing concentrations of nitrate, chlorophyll, and oxygen in the Gulf and look for patterns that suggest connections among the three variables. Students observe catch-per-unit-effort values for shrimp catches and determine whether or not these data support the hypothesis that an increase in nitrogen use in the Mississippi Basin adversely affects shrimp harvests in the Gulf. Finally, students put the whole story together, synthesizing concepts and explaining interactions among the Earth's spheres.

Instructional Context

This is an opportunity for students to investigate an issue of topical interest that is not yet conclusive. They analyze several forms of data to evaluate a current scientific hypothesis.

What Does the Ocean Floor Look Like?

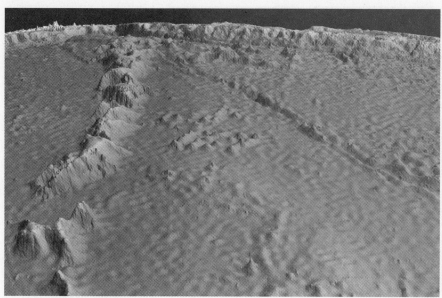

Jennifer Loomis, TERC

Investigation Overview

Students observe a map showing global relief and hypothesize about the shape of the ocean floor and the location of specific features. They view animations of water draining from the Atlantic and Pacific Oceans, revealing the topography of the ocean floor. Using images of the drained oceans, students identify specific seafloor features and infer how they were formed. Students explore differences between active and passive continental margins.

Instructional Context

This investigation can be used as an introduction to ocean floor features. It provides students an opportunity to visualize the unfamiliar rocky surface of the 70% of our planet that is covered by oceans. The investigation also serves as a review of plate tectonic concepts, as most of the ocean floor features encountered are formed by plate tectonic processes.

KEY CONCEPTS

- The geological features of the ocean floor are formed by tectonic activity.
- Passive continental margins are associated with wide continental shelves.
- Active continental margins are associated with steep continental slopes, trenches, and tectonic activity.

KEY SKILLS

- Interpreting bathymetric maps
- Identifying ocean floor features
- Inferring how ocean floor features were formed

ESTIMATED TIME REQUIRED

- 45 minutes Internet use

Teacher's Guide Chapter 23
Internet Investigation

When Were the Atlantic and Pacific Oceans Separated by Land?

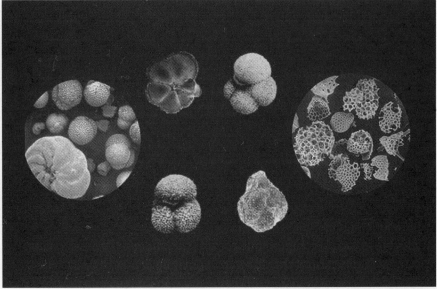

Brian Huber, Smithsonian Institute

Copyright © McDougal Littell Inc.

KEY CONCEPTS

- Microfossils record changes in ocean chemistry.
- Formation of the Isthmus of Panama affected ocean chemistry and climate on both sides of it.

KEY SKILLS

- Interpreting salinity and temperature data
- Identifying microfossils in sediment samples
- Making inferences from multiple sources of data

ESTIMATED TIME REQUIRED

- 30 minutes Internet use
- 10 minutes deskwork or homework

Investigation Overview

Students examine the salinity and temperature of the Caribbean Sea and the Pacific Ocean on either side of the Isthmus of Panama. They describe how atmospheric processes control ocean chemistry on either side of the isthmus, then analyze trends in oxygen isotope data from microfossils to show that the two seas once shared a similar chemistry. Students watch an animation of plate motion that formed a land bridge between North and South America. To date the formation of the land bridge, students analyze the distribution of a fossil species, Pullentiatina, and interpret data to determine the date that this microorganism appeared only on the Pacific Ocean side of the isthmus. By dating this evolutionary change, students estimate the date of the formation of the Isthmus of Panama. Finally, students discuss the interactions among the geosphere, hydrosphere, atmosphere, and biosphere before and after formation of the isthmus.

Instructional Context

To introduce the investigations, you might ask students to imagine how different the world's oceans might be if the Isthmus of Panama didn't exist. Water moving freely between the Atlantic and Pacific Oceans would result in a single environment, and marine life would be distributed equally throughout the channel. Next, ask students to think about the differences in ocean conditions that do exist on either side of the isthmus. Consider how a seafloor might preserve a record of the type of ocean conditions above it. Prior to the investigation, students should understand how tectonic activity can build landmasses and that salinity varies throughout the ocean.

How Can One Ocean Current Affect the Whole North Atlantic?

NOAA Historic C&GS Collection

KEY CONCEPTS

• Warm water transported by the Gulf Stream affects the climate of coastal northern Europe.
• The Gulf Stream forms eddies that can be productive fishing areas.

KEY SKILLS

• Observing patterns in temperature data
• Inferring relationships between sea temperatures and phytoplankton concentration
• Comparing and interpreting imagery depicting the Atlantic Ocean

ESTIMATED TIME REQUIRED

• 45 minutes Internet use

Investigation Overview

Students examine differences in winter temperatures for cities at the same latitude and identify a pattern: coastal northern Europe has warmer winter temperatures than Canada or inland Europe. Students explore images of sea surface temperatures (SST) in the Atlantic Ocean and relate the warm waters of the Gulf Stream to the pattern of winter temperatures they observed. Students view animations showing the formation of warm and cold eddies in the Gulf Stream and identify each type in SST images. Finally, students explore the connection between colder sea temperatures and the abundance of phytoplankton; students infer that cold core eddies in the Gulf Stream can be productive fishing grounds.

Instructional Context

The Gulf Stream is one of Earth's most prominent ocean currents. It is often compared to a huge river in the ocean, and it shares some similarities with a jet stream in the atmosphere. Benjamin Franklin originally mapped the general location of the Gulf Stream using data from ships. Today, instruments on satellites monitor the world's oceans, and the daily location and conditions of the Gulf Stream are broadcast to ships, weather forecasters, and people in the fishing industry.

How Do Tides Work?

Carl Baldwin, NOAA

KEY CONCEPTS

- The relative positions of the moon, sun, and Earth control tide levels.
- Two high tides and two low tides occur every 24 hours and 50 minutes.

KEY SKILLS

- Determining relationships between tides and lunar phases
- Interpreting tide charts
- Predicting tide patterns

ESTIMATED TIME REQUIRED

- 35 minutes Internet use

Investigation Overview

Students predict tide patterns for two specific days, one week apart, based on images of the moon's phases. They observe a sequence of still images showing tide levels at Cape Porpoise, Maine, for these two days and make inferences about the relationships between tides and moon phases. Students observe time-lapse animations of changing tide levels correlated with corresponding data on graphical tide charts. Students interpret the visualizations to compare tide patterns and predict when the next high or low tide will occur. Students also predict tide patterns for another day of the month, when the moon is full.

Instructional Context

This investigation gives students the opportunity to analyze changes in tidal levels from time sequence photographs. The visual changes are linked with a diagram that shows how tidal changes are depicted graphically. Students should already be familiar with material in Chapter 24 about the alignment of the sun, moon, and Earth during various phases of the moon.

Could You Break the Record for an Ocean Sailboat Race?

Arttoday.com

KEY CONCEPTS

- Journeys across the oceans require a study of ocean phenomena.
- Ocean phenomena can be observed, predicted, and used to one's advantage to avoid potential dangers or obstacles.

KEY SKILLS

- Identifying ocean phenomena that are relevant to a global sailing voyage
- Interpreting ocean data
- Synthesizing observations

ESTIMATED TIME REQUIRED

- 35 minutes Internet use
- 15 minutes deskwork or homework

Investigation Overview

Students make an initial prediction of the best route to take for breaking the record for a sailboat race around the world. They examine phenomena such as global winds, ocean currents, and weather conditions that would affect their journey. Students consider these features as advantages, limiting factors, or obstacles to their goal of sailing around the world. After completing the online investigation, students revisit their maps and draw a new or revised route, incorporating their new understandings, and they write a detailed description of the route they would follow.

Instructional Context

This is a culminating activity that allows students to draw on their knowledge of many concepts about ocean currents and winds. The challenge requires students to synthesize understandings about interactions among Earth's spheres that affect movement of the oceans.

Teacher's Guide Chapter 24
Internet Investigation

US Army Corps of Engineers

Teacher's Guide Unit 6
Internet Investigation

Investigation Overview

Students observe a variety of visualizations to develop an understanding of ocean temperatures and circulation patterns that characterize El Niño and La Niña events. They explore graphic representations of sea surface temperature, ocean buoy data, and movie clips of satellite-derived imagery. Next, students analyze maps illustrating changes in atmospheric circulation and weather patterns during El Niño-La Niña events. Students note how weather patterns in their own regions are affected by El Niño and La Niña circulation patterns. The investigation culminates with students developing a presentation on one weather-related impact of El Niño-La Niña.

Instructional Context

El Niño-La Niña ocean temperature and atmospheric circulation patterns influence the entire Earth system through extreme weather events, including severe drought, powerful coastal storms, increased tornado activity, and heavy snows. Dramatic shifts in temperature and precipitation patterns around the world are results of El Niño and La Niña patterns. Sophisticated instruments on ocean buoys and satellites are capable of measuring small changes in sea temperature and sea surface height. These data are used to study the El Niño and La Niña processes. The intensity and frequency of El Niño and La Niña events appear to be increasing, perhaps in response to global climate change. Atmospheric circulation patterns, especially the average location and intensity of jet streams, are altered while El Niños and La Niñas are occurring.

KEY CONCEPTS

- El Niño and La Niña are names applied to specific ocean temperature and circulation patterns in the Pacific Ocean.
- El Niño-La Niña patterns strongly influence the global atmospheric system, sometimes causing extreme weather phenomena.

KEY SKILLS

- Analyzing animations, satellite images, and maps
- Identifying the stages of El Niño and La Niña events
- Summarizing information

ESTIMATED TIME REQUIRED

- 90 minutes Internet use
- 60 minutes deskwork or homework

Can We Blame El Niño for Wild Weather? ▶ ESU601

Unit 6 Project Description

How does El Niño-La Niña affect the Earth system? Using Web resources, students identify, explore, and research an impact caused by El Niño or La Niña ocean circulation patterns. Students develop a presentation that explains the impact and how it affects the Earth system. They present their findings in a format that you have approved.

Criteria for setting student expectations and grading the project are provided below. In conjunction with the criteria, you may also want to use the assessment rubric to evaluate student projects. The rubric is found in the Introduction of the Teacher's Guide.

Unit Project Criteria

1. **A description of the El Niño- or La Niña-related impact. (10%)**
   - Explain the impact, its location, and timing.
   - Use images, graphics, or animations to illustrate the impact.
2. **An explanation of the cause-and-effect relationship between El Niño-La Niña and the impact. (20%)**
   - Describe how El Niño-La Niña is thought to cause the impact.
   - Use images, graphics, or animations to illustrate the cause-and-effect relationship.
3. **An explanation of the scientific evidence for the cause-and-effect relationship. (25%)**
   - Interpret the scientific evidence, how it was gathered, and how it provides evidence supporting the cause-and-effect relationship.
   - Identify strengths and any uncertainties in the scientific evidence.
4. **A description of Earth system interactions involved in this cause-and-effect relationship. (25%)**
   - Identify and describe interactions in the Earth system related to the impact.
   - Illustrate interactions with images, graphics, or animations.
5. **A presentation of research and results. (20%)**
   - Communicate findings and ideas clearly and accurately.
   - Present appropriate data.
   - Use appealing visuals and design.

Alternative Projects

Students may want to choose their own research topic or pursue one of the Questions for Discovery suggested below. The assessment rubric in the Teacher's Guide provides a standard basis for evaluating student performance and is recommended for grading alternative projects.

Sample Questions for Discovery

- What additional scientific research should be conducted for better understanding of the causes of El Niño-La Niña and its impacts?
- What is the relationship between El Niño-La Niña and global climate change?
- Using historical and current data, predict when the next El Niño will occur.

7 UNIT Space

Visit Mars from your desktop
Search for signs of water on Mars. Choose a landing site where you would look for signs of life.

Model asteroid impacts on Earth and the moon
Choose the size, speed, and density of asteroids that might collide with Earth and the moon. View examples of craters formed on each body.

Find how sunspots affect communications on Earth
Examine sunspots and explore how they not only cause beautiful auroras but also disrupt satellite communications on Earth.

NASA

Unit 7 Space

7 UNIT Space

OTHER WEB RESOURCES

VISUALIZATIONS

Spark new ideas for investigations with images and animations, such as:

• Solar System

• Impact theory of moon formation

• Radar mapping of Venus

• Life stages of stars

DATA CENTERS

Extend your investigations with current and archived data and images, such as:

• Mars exploration

• Remotely-accessed telescopes

• Night-time sky

• Meteorites

EARTH SCIENCE NEWS

Relate your investigations to current events around world, such as:

• Missions to Mars

• Search for extra-solar planets

• Solar and lunar eclipses

• New discoveries

Unit 7
Space

What if Earth and the Moon Were Hit by Twin Asteroids?

NASA/JPL

Investigation Overview

Students are asked to consider the hypothetical situation of identical asteroids with the same mass and speed striking both Earth and the moon. Students predict how the resulting craters would compare, and consider how each crater would look after ten million years. Students examine lunar and terrestrial craters of various sizes and explore their similarities and differences. Students discover that Earth has been hit by more asteroids than the moon has, but surface processes on Earth actively erase the evidence. Finally, students use the Impact Calculator tool to find out how an asteroid's speed, diameter, angle of impact, and composition affect crater size and structure on Earth and the moon.

Instructional Context

The Impact Calculator tool gives a good approximation of crater size and structure for projectiles larger than about one kilometer in diameter. The Impact Calculator was developed with the assistance of Professor H. Jay Melosh of the University of Arizona, an expert on impact cratering events. If you wish to extend this investigation to examine cratering on other planets, similar calculators are available on the Internet that allow students to model impacts on a variety of planetary bodies.

KEY CONCEPTS

- Earth and the moon have experienced a comparable number of large asteroid impacts.
- Processes on Earth's surface erase impact craters.
- Impact crater size depends mostly on the kinetic energy of the projectile.

KEY SKILLS

- Interpreting the relative scarcity of terrestrial impact craters
- Calculating crater sizes based on speed, diameter, angle, and composition of the projectile
- Comparing the influence of projectile properties on crater dimensions and structure

ESTIMATED TIME REQUIRED

- 45 minutes Internet use

Teacher's Guide Chapter 25 Internet Investigation

Why Does the Size of the Sun Appear to Change? ▶ ES2603

Larry Kendall, modified from NASA/GSFC

Investigation Overview

Students watch an animation of the sun at monthly intervals for one year and measure the apparent diameter of the sun. From the diameters, they calculate and plot the Earth-sun distance. Using the plotted orbit, students determine the times of perihelion and aphelion and compare orbital eccentricities of the Earth with other planets. Finally, students make inferences about the relationship between Earth's orbit and the seasons.

Instructional Context

You may wish to have students plot their Earth orbits on a large sheet of paper instead of on the Web site, for more accuracy. Draw lines radiating from the center of the paper every 30 degrees and use a scale for the distance, such as 1 cm = 10 million kilometers.

The last Web page of the investigation is designed to counter the common misconception that Earth's orbit, rather than the tilt of its axis, is the cause of the seasons. In fact, about 7% more solar energy strikes Earth at perihelion (January) than at aphelion (July). Students can see this by finding the apparent area of the solar disk at perihelion and aphelion in square pixels, then calculating the ratio of the two areas. They should get a ratio of about 1.07.

KEY CONCEPTS
- Changes in the apparent size of the sun are due to Earth's elliptical orbit.
- The sun appears largest at perihelion and smallest at aphelion.
- Earth's non-circular orbit is not the cause of the seasons.

KEY SKILLS
- Measuring the apparent diameter of the sun
- Plotting Earth-sun distance to determine shape of Earth's orbit
- Analyzing plot to estimate perihelion and aphelion dates

ESTIMATED TIME REQUIRED
- 40 minutes Internet use

Teacher's Guide Chapter 26
Internet Investigation

How Does the Sunspot Cycle Affect Earth? ▶ ES2605

Larry Kendall, modified from NASA/GSFC/NOAA/USGS

Investigation Overview

In this investigation, students examine animations of the sun at visible and x-ray wavelengths to see solar activity related to sunspots. Students explore short-term solar effects on Earth and interpret graphs of sunspot patterns. Finally, students compare long-term graphs of sunspot numbers and terrestrial phenomena to look for possible relationships.

Instructional Context

Since the invention of the telescope, people have watched and recorded sunspots and wondered what effect, if any, they have on Earth and its inhabitants. In the past decade, our ability to study and predict "solar weather" related to sunspots has increased dramatically.

Points to emphasize during the investigation are: solar weather events are not new, but they become more relevant as our dependence on satellites and other vulnerable technologies grows; understanding and planning for appropriate protection from these effects have important economic benefits; and cause-and-effect relationships between solar activity and long-term phenomena such as climate are difficult to establish.

Though it is not a prerequisite, students will better appreciate solar effects if they have a general understanding of electricity and magnetism. As an ongoing activity, have students check the daily solar weather report on the Web at NOAA's Space Environment Center, and watch for related current events such as aurorae and solar-induced power grid failures

How Fast Does the Wind Blow on Jupiter? ▶ ES2704

NASA/JPL/University of Arizona

KEY CONCEPTS

- Studying other planets can increase our understanding of Earth.
- Planetary wind speeds can be measured by analyzing the motion of features embedded in global wind belts.

KEY SKILLS

- Observing and comparing global winds on Earth and Jupiter
- Interpreting global wind patterns
- Measuring distances and calculating wind speeds

ESTIMATED TIME REQUIRED

- 30 minutes Internet use

Investigation Overview

Students view movies of cloud motion on Earth and Jupiter, comparing global wind circulation on the two planets. Using a detailed animation of Jupiter's atmosphere taken over thirty-five Earth days, students measure the distance traveled by selected features at six different latitudes. Students calculate and graph the wind speed and direction at each latitude, then relate wind speeds to Jupiter's belts and zones. Finally, students compare the speed of winds on Jupiter to wind speeds on Earth and make inferences about the reasons for the differences they observe.

Instructional Context

Other planets can serve as natural laboratories for studying planetary atmospheres. Unlike the inner planets, the gas giants—Jupiter, Saturn, Uranus, and Neptune—don't have solid surfaces. Because surface features serve to define a planet's rotation, it's difficult to determine the relative speed and direction of atmospheric circulation for these planets. Studies of Jupiter's clouds and magnetic field show that the Great Red Spot is fixed relative to the planet's interior, allowing us to use it as a reference point for making measurements. Students should be familiar with Earth's global atmospheric circulation (Chapter 19) before they begin this investigation.

What Processes Shape Planetary Surfaces?   ▶ **ES2708**

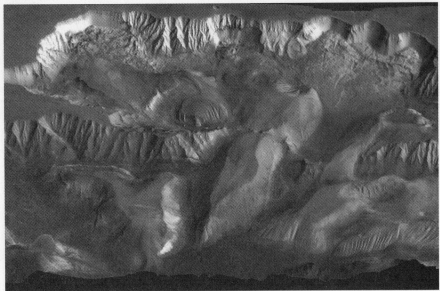

NASA/GSFC/JPL

KEY CONCEPTS

- The processes that shape other planetary bodies are the same ones that shape Earth.
- Planetary surface processes can be inferred from the features they produce.

KEY SKILLS

- Observing and interpreting planetary features
- Inferring processes from features

ESTIMATED TIME REQUIRED

- 30 minutes Internet use

Investigation Overview

Students begin the investigation by considering the history of solar system exploration. Next, they view images of the surfaces of Earth, each of the inner planets, and several large moons. Students examine surface features in these images and interpret the processes responsible for forming them. Students view "before" and "after" images of Mars and of Jupiter's moon Io to see how their surfaces have changed over time and make inferences about the processes responsible for the changes. Finally, students consider the processes that are unique to Earth.

Instructional Context

Most planetary surface processes occur over long periods of time or as the result of cataclysmic events. We study Earth's features and the processes that form them, and use this knowledge to interpret the origin of similar features on other planets. The gaseous outer planets lack solid surfaces, but their large moons are essentially planets whose crusts are sculpted by similar processes.

What Does the Spectrum of a Star Tell Us about Its Temperature?

NASA

KEY CONCEPTS

- Patterns in stellar spectra can be used to infer stars' temperatures.

KEY SKILLS

- Analyzing and categorizing stellar spectra
- Determining stellar temperatures from simulated spectra

ESTIMATED TIME REQUIRED

- 30 minutes Internet use

Investigation Overview

Students examine real and simulated stellar spectra. Based on the width and location of lines, students devise a classification scheme for similar spectra. After they produce their own categories, students sort the spectra according to models that represent standard spectral types. Students then examine peak emission wavelength data—the wavelength of highest energy output—for each spectrum. They discover that spectra within each standard class have similar peak emission wavelengths. Students continue their analysis by using the star's peak wavelength to infer its temperature. Students finish their investigation by examining simulated spectra of four "unknown" stars to determine their temperatures.

Instructional Context

Astronomers analyze the unique pattern of lines in stellar spectra to deduce the elemental composition and temperature of stars. The simulated spectra in the investigation give students experience analyzing patterns of lines in order to group them into spectral types. You may wish to build an analogy relating spectra to Universal Pricing Code (UPC) symbols, or barcodes. Stars within the same spectral class could be likened to grocery items within the same department—the barcodes for items within a single department would all be similar to one another, yet each code has some unique characteristics.

Teacher's Guide Chapter 28
Internet Investigation

What Happens as a Star Runs Out of Hydrogen? ▶ ES2810

NASA/ESA/Hubble Heritage Team (STScl/AURA)

KEY CONCEPTS

- Hydrogen is the primary atomic fuel for fusion reactions in stars.
- The larger a star's mass, the faster it uses its fuel.
- The final state of stellar matter is dependent on the star's initial mass.

KEY SKILLS

- Interpreting diagrams of processes in stellar interiors
- Inferring the final state of stars from their initial masses.

ESTIMATED TIME REQUIRED

- 30 minutes Internet use

Investigation Overview

Students explore an interactive animation depicting what happens to stars of three different masses as they run out of hydrogen in their nuclear-fusing cores. Students observe the three stars pass through different stages of their life cycles, including phases such as supernova, red giant, planetary nebula, and white dwarf. The animation allows students to observe changes in color, brightness, and size for each type of star over time. They can also overlay a diagram on the animation that depicts processes inside the stars at each phase in their evolution.

Instructional Context

Stellar processes can be presented as an ongoing battle between the inward collapsing force of gravity and the outward pressure generated by nuclear fusion. Each phase of stellar evolution can be classified according to which force is more dominant at the moment. Students can make their own timelines to illustrate what happens as a star runs out of hydrogen. Access images of stars from the Hubble Space Telescope's Website or from calendars or magazines. Have students place the images on a timeline to indicate the sequence of stellar death.

Could Mars Support Life?

NASA/JPL/Malin Space Science Systems

KEY CONCEPTS

- Comparisons between Mars and Earth help scientists design strategies for exploring Mars.
- Life thrives in a wide range of conditions on Earth.
- Life may be present on other worlds.
- Mars appears to have had liquid water on its surface in the past.

KEY SKILLS

- Interpreting satellite images
- Providing evidence to support a conclusion

ESTIMATED TIME REQUIRED

- 60 minutes Internet use
- 60 minutes deskwork or homework

Investigation Overview

Students begin by comparing Earth and Mars from space and at the surface. They consider the requirements for life, based on the range of life on Earth. Focusing on liquid water as a key requirement, students compare satellite views of Earth and Mars to look for evidence of water. Next, students explore eight sites on Mars. They describe the unique aspects of each site and decide if the sites show evidence of flowing water and, thus, the potential for supporting life. Students consider which of the sites might be a good place to land a spacecraft to search for evidence of past or current life. Based on the data they collect, students prepare a presentation to argue a case either for or against the likelihood that Mars has supported, or could support, life. Finally, students review the status of current missions to Mars and any new discoveries in the search for life on Mars.

Instructional Context

The prospect of life on other worlds has long been a source of fascination for humans. Recently, the search for extraterrestrial life has taken on increasing depth and sophistication. Scientists searching for life assume that liquid water is a requirement for life; therefore they concentrate on areas that show evidence of water-related processes. Images from spacecraft orbiting Mars show channels, streamlined islands, and sedimentary layers that may have formed during periods when water was present on Mars' surface. A comprehensive series of missions to Mars over the next decade will provide new data for continuing the search for life on Mars.

Unit 7 Project Description

Using the results of this investigation and Web-based research on current missions to Mars, students prepare a presentation to argue a case either for or against the likelihood that Mars has supported, or could support, life. The case they make must be based on data and research. They present their findings in a format that you have approved.

Criteria for setting student expectations and grading the project are provided below. In conjunction with the criteria, you may also want to use the assessment rubric to evaluate student projects. The rubric is found in the Introduction of the Teacher's Guide.

Unit Project Criteria

1. **A description of the requirements for life. (25%)**
   - Describe the requirements for life on Earth.
   - Explain what life in extreme habitats on Earth tells you about the possibility of life on Mars.
   - Describe similarities and differences between Mars and Earth related to their abilities to support life.

2. **An evaluation of the data from the eight sites on Mars. (40%)**
   - Summarize your observations.
   - Evaluate the evidence of flowing water or other signs related to the potential for life.

3. **An explanation of the findings of your Web-based research. (20%)**
   - Summarize your findings from recent missions to Mars.
   - Use images, graphics, or animations to illustrate the findings.

4. **An effective presentation of your conclusions. (15%)**
   - Communicate findings and ideas clearly and accurately.
   - Present appropriate data that support your conclusions.
   - Use appealing visuals and design.

Alternative Projects

Students may want to choose their own research topic or pursue one of the Questions for Discovery suggested below. The assessment rubric in the Teacher's Guide provides a standard basis for evaluating student performance and is recommended for grading alternative projects.

Sample Questions for Discovery

- How have advances in technology influenced the search for life on Mars?
- Could Europa support life?
- How might we search for life on recently-discovered extrasolar planets.

Teacher's Guide Unit 7 Internet Investigation

8 UNIT Earth's History

Jennifer Loomis, TERC

Explore the Grand Canyon

Read the rock record of environmental change revealed by layers of the Grand Canyon.

Use fossils to investigate how life has changed

Compare fossils from different periods of Earth's history to explore how life has changed over time.

Investigate the causes of mass extinctions

Explore evidence from the rock record to look for the causes of mass extinction events.

Unit 8
Earth's History

Earth Science

UNIT 8 Earth's History

OTHER WEB RESOURCES

VISUALIZATIONS

Spark new ideas for investigations with images and animations, such as:

• Events thru geologic time

• How fossils form

• Break-up of Pangea

• Asteroid impacts

DATA CENTERS

Extend your investigations with current and archived data and images, such as:

• Determining the age of rocks

• Grand Canyon case study

• Paleontology

• Reading history in sediments

EARTH SCIENCE NEWS

Relate your investigations to current events around world, such as:

• Fossil finds

• Endangered species

• New discoveries and insights

Unit 8
Earth's History

Copyright © McDougal Littell Inc.

What Stories Do Rocks Tell?

Arttoday.com

Investigation Overview

Students examine photographs and drawings, both line and block, of rock sequences. They use a series of animations to explore geologic processes and interpret the relative ages of the rock layers. Students observe deposition of layers, faulting and folding of layers, and the combined processes that result in an unconformity. At the end of the investigation, students are presented with a new site to interpret. They number the rock layers from oldest to youngest and write a detailed description of the geologic history of the site.

Instructional Context

The Colorado Plateau in the southwestern United States is a geologically rich area of the country. Many national parks, including Bryce, Canyonlands, Arches, Zion, and the Grand Canyon, are located in this region. Consider having students prepare research reports or presentations focused on the geologic story of each of the different parks or monuments in this area of the southwest. Encourage them to include labeled photographs or diagrams portraying stratigraphic sequences.

You may also wish to call students' attention to the fact that one of the best ways to learn about the geologic history of an area is by examining roadcuts. Suggest that when they travel they make sketches or take photographs of exposed rock sequences at the sides of roads.

KEY CONCEPTS

- Using relative dating principles, the sequence of geologic events that occurred at a site can be deduced.
- In an undeformed sequence of sedimentary rocks, the oldest rocks are at the bottom.
- Intrusions and faults are younger than the rocks they cut.
- Tilted, folded, or faulted layers of sediments were originally horizontal.

KEY SKILLS

- Interpreting stratigraphic sequences
- Ordering geologic events
- Reconstructing geologic history

ESTIMATED TIME REQUIRED

- 35 minutes Internet use

Teacher's Guide   Chapter 29
Internet Investigation

How Do Trees Record Time?

Tom Grace, TERC

KEY CONCEPTS

- Some trees produce annual growth rings that can be used to date environmental events.
- Skeleton plotting allows ring widths to be standardized for comparison across samples.
- Tree rings can be used to measure absolute time.

KEY SKILLS

- Interpreting tree ring data
- Correlating climatic events with tree rings
- Synthesizing patterns

ESTIMATED TIME REQUIRED

- 45 minutes Internet use

Investigation Overview

In this simulation of tree ring cross-dating, students correlate climatic events recorded in tree rings to date a prehistoric dwelling. Students look for "wet" years and "dry" years by comparing tree ring widths. They use skeleton plotting, a technique to standardize tree ring width comparisons, and build a master chronology by pattern matching. Then, they pinpoint the dates of key events and calculate the ages of additional samples.

Instructional Context

Tree ring chronologies extend for several thousand years into the past, providing an absolute time scale useful for dating archeological sites and examining climate patterns. Tree ring dating, or dendrochronology, is so accurate that it can be used to calibrate other dating methods. Provide samples of cut tree rounds to students in class and ask them to identify "wet" and "dry" years recorded by the tree. Discuss how annual rings would appear in trees from landscaped areas that are watered regularly every year.

How Did the Layers of the Grand Canyon Form?   ▶ ES2906

NPS

KEY CONCEPTS
- Rocks provide clues about the environments in which they formed.
- The geologic history of an area can be deduced from the rocks that formed there.

KEY SKILLS
- Examining rock features for clues about their formation
- Inferring the sequence of past environments from sedimentary rocks

ESTIMATED TIME REQUIRED
- 45 minutes Internet use

Investigation Overview

Students examine a cross-sectional diagram and photographs of the major rock units exposed in the Grand Canyon. They analyze rock features and fossils for clues about the environment in which the rock was formed. Beginning at the bottom of the canyon, students explore each rock unit to deduce the sequence of environmental changes that occurred in this region over time.

Instructional Context

The Grand Canyon offers one of the most spectacular views of undisturbed sedimentary layers on Earth. The sedimentary layers in any geologic sequence directly reflect the environments in which they formed. After students practice reading the environmental clues from rock layers in the Grand Canyon, they will be aware of what to look for when examining and interpreting the geologic history of any area. You may wish to extend the investigation by having students interpret rock layers in your own region, or conduct research on how the Colorado River cut into the sequence of rocks to form the Grand Canyon.

*Teacher's Guide Chapter 29
Internet Investigation*

How Has Life Changed over Geologic Time?

▶ **ES3002**

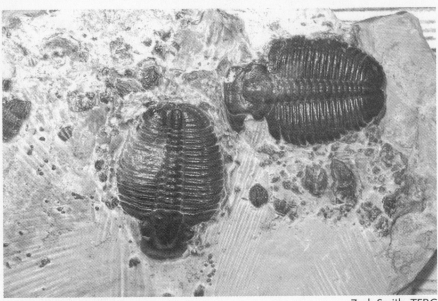

Zach Smith, TERC

Investigation Overview

Students begin by exploring images of fossils. They are introduced to the concept of fossils as indicators of environmental condition. Next they take a geologic journey to up to nine fossil sites. At each site they conduct research to discover the types of organisms that once lived there, as well as the conditions that were present during the time frame in which they lived. Teams or the entire class may need to share information to complete their tables. Then, students investigate evolutionary patterns as they drag and drop symbols for life forms onto a geologic time scale. Last, students pick a favorite fossil and write a paragraph about it.

Instructional Context

If you have access to fossil specimens, consider sharing them with students prior to the investigation. Encourage students to think about the types of rocks that commonly contain fossils. Chapter 30 in the textbook presents different types of organisms that lived during each geologic time period. This sequence is the primary concept students discover through this investigation. If you plan to use both the investigation and the textbook to cover this concept, the textbook reading will be more meaningful to students after they complete the investigation.

Where and When Did Dinosaurs Live?

Tom Grace, TERC, modified from USGS

Investigation Overview

For over 100 million years, dinosaurs roamed Earth. During that time span a wide variety of dinosaurs evolved and went extinct. Evidence of their existence is preserved in the fossil record. In this investigation, students conduct a case study of a single type of dinosaur. They prepare a "Day in the Life of ....." report and then compare their findings with those of students who examined other dinosaurs. They then compile information from the different species in order to construct a dinosaur timeline.

Instructional Context

Students are often familiar with the names of many dinosaurs. However, they frequently hold mistaken notions about the behavior of individual dinosaurs and the time spans during which they lived. For instance, a common misconception held by students and perpetuated by movies is that *Stegosaurus* and *Tyrannosaurus rex* could have engaged in battle. In fact, these two never would have met, as *Stegosaurus* went extinct about 80 million years before *Tyrannosaurus rex* came along. Help students recognize the vast spans of geologic time during which dinosaurs existed.

Some creatures are mistakenly referred to as dinosaurs. The ichthyosaur and the pteranodon are two such beings. Ichthyosaurs were marine organisms resembling modern-day dolphins. Pteranodons were flying reptiles; they are included here because many students are familiar with them. As they conduct research, students should discover that pteranodons are *not* dinosaurs. However, be sure to discuss this information with them at the end of the investigation. It may also be useful to lead a culminating class discussion based on the question "What have you learned about dinosaurs that is different from what you thought about dinosaurs *before* you conducted your case study?"

KEY CONCEPTS

- Many different dinosaurs have existed on Earth, but not all at the same time.
- Skeletal remains, footprints, and other evidence provide clues to dinosaur behavior and the conditions existing on Earth during the time they lived.

KEY SKILLS

- Identifying basic characteristics of dinosaurs
- Categorizing dinosaurs by geologic time span
- Appreciating the great lengths of geologic time during which dinosaurs existed

ESTIMATED TIME REQUIRED

- 45 minutes Internet use
- 45 minutes deskwork or homework

**Teacher's Guide   Chapter 30
Internet Investigation**

What Caused the Mass Extinction Recorded at the K-T Boundary?

Don Davis, NASA

Key Concepts

- A mass extinction of organisms occurred about 65 million years ago.
- Mass extinctions have occurred periodically throughout Earth's history.
- Scientific theories are built from explanations of the available evidence.

Key Skills

- Analyzing evidence
- Evaluating credible hypotheses
- Drawing conclusions

Estimated Time Required

- 90 minutes Internet use
- 90 minutes deskwork or homework

Investigation Overview

Students begin this investigation by exploring data from rock layers deposited at the Cretaceous-Tertiary (K-T) boundary. They observe images and analyze fossil counts. Next, students research the theories that attempt to explain the events. Students consider how the K-T extinction compares to other mass extinctions throughout Earth's history. At the end of the investigation, students will prepare a presentation to share their research findings. In these projects, students build a case to support an existing theory, or formulate a new theory, that explains the cause of a mass extinction event.

Instructional Context

Sixty-five million years ago, at the boundary between the Cretaceous and the Tertiary (K-T) periods, more than half of all the species of plants and animals living on Earth became extinct. Whatever occurred at the K-T boundary resulted in widespread environmental change on Earth. Many students may already be aware of the theory that an asteroid impact led to the extinction of dinosaurs. Though the asteroid impact theory is widely accepted today, there is still debate on the role that volcanism may have played. Furthermore, three of the largest volcanic basalt floods correlate with mass extinctions. Recent research has revealed that the cause of the earlier Permian extinction may also have been extraterrestrial in origin. Help students consider all credible theories as they conduct their research.

What Caused the Mass Extinction Recorded at the K-T Boundary?

Unit 8 Project Description

Using the results of this investigation and Web-based research on current and new discoveries about mass extinctions, students prepare a presentation that builds a case to support an existing theory, or that formulates a new theory, explaining the cause of a mass extinction. The case they make must be based on data and research. They present their findings in a format that you have approved.

Criteria for setting student expectations and grading the project are provided below. In conjunction with the criteria, you may also want to use the assessment rubric to evaluate student projects. The rubric is found in the Introduction of the Teacher's Guide.

Unit Project Criteria

1. **A description of the location where the extinction event occurred. (10%)**
   - Prepare a map showing specific location(s) involved with the extinction event.
   - Cite evidence to support the location(s) you identified for the extinction event.

2. **An explanation of when the extinction event occurred. (10%)**
   - Tell how many years ago and in which evolutionary period the event occurred.
   - Cite evidence that allows you to date the event.

3. **An explanation of the evidence you used to make your case. (20%)**
   - Explain the source from which your evidence comes (geological, biological, extraterrestrial).
   - Explain clearly how you interpret the evidence.

4. **A summary of your theory of how the extinction occurred. (40%)**
   - Explain how the evidence supports this.
   - Explain the story in terms of interactions among Earth's spheres.

5. **An effective presentation of your conclusions. (20%)**
   - Communicate findings and ideas clearly and accurately.
   - Present appropriate data that support your conclusions.
   - Use appealing visuals and design.

Alternative Projects

Students may want to choose their own research topic or pursue one of the Questions for Discovery suggested below. The assessment rubric in the Teacher's Guide provides a standard basis for evaluating student performance and is recommended for grading alternative projects.

Sample Questions for Discovery

- What caused the Permian extinction?
- How do the frequency and timing of impact cratering events compare to the frequency and timing of mass extinctions?
- How does the global distribution of impact craters compare to the global distribution of basalt flows?

Internet
Investigation Guide

Face of the Earth™ ARC Science Simulations Copyright © 2001 www.arcscience.com

Answered
Student Worksheets

1 UNIT Investigating Earth

NASA

Investigate Earth from space

See Earth in new ways. Interpret photographs taken by astronauts and analyze images made by satellite instruments.

Ask questions and design investigations

Work with the same data and tools scientists use to investigate the Earth system.

Interact with 3-D models of landforms

Rotate models of mountains, cliffs, and other features to read topographic maps.

Unit 1
Investigating Earth

1 UNIT Investigating Earth

OTHER WEB RESOURCES

VISUALIZATIONS

Spark new ideas for investigations with images and animations, such as:

- Remotely-sensed images
- Earth's interacting spheres
- Paths of the water cycle
- One place at many scales

DATA CENTERS

Extend your investigations with current and archived data and images, such as:

- Earth from space
- Earth as a system
- Interactive Earth models
- Earth mapping

EARTH SCIENCE NEWS

Relate your investigations to current events around world, such as:

- Environmental news
- Satellite launches
- Earth science discoveries

Unit 1
Investigating Earth

1 List several different parts of the Earth system you can see in the image.

Answers will vary. Land, water, clouds, ice, plants, and air are possible answers.

2 List Earth's four spheres. Give several examples of features in each sphere.

Geosphere: Rocks, mountains, dirt, metals

Hydrosphere: Oceans, rivers, lakes, groundwater, rain, ice, snow

Atmosphere: oxygen, nitrogen, water vapor

Biosphere: humans, plants, fish, mammals, bacteria

3 Why do you think rain and snow are listed with the hydrosphere instead of the atmosphere?

Rain and snow are forms of water. When water vapor in the atmosphere condenses to a liquid or solid, it is part of the hydrosphere.

4 For each image:
- List the major features and tell which sphere each one represents.
- Describe how each feature shows interactions between the spheres.
- Whenever possible, follow the results of an interaction through all four spheres.

Answers will vary. These are sample responses.

A. Suez Canal

The canal is cut through the geosphere (land). It shows that the biosphere (humans) changed the geosphere (land) to connect two parts of the hydrosphere (ocean).

B. Tornado

Atmosphere (wind) brings hydrosphere (rain) and damages biosphere (trees) and buildings (biosphere and geosphere materials).

C. Tropical Island

Where geosphere (land) is above the hydrosphere (sea), the biosphere (plants) is able to thrive.

4 continued

D. Forest fire

Burning of the biosphere (forest) is adding energy and materials to the atmosphere and returning materials to the geosphere (land).

E. Oil wells burning

"Fossilized" biosphere material (oil) drilled from the geosphere (land) reacts with the atmosphere, releasing energy and materials into the atmosphere

F. Amu Darya River

Biosphere(humans) control the hydrosphere (river) to nurture biosphere (crops) causing transpiration of water vapor into the atmosphere.

G. Wind farm

Biosphere (humans) use geosphere materials (metal, plastics) to harness energy from the atmosphere.

H. Mount Etna, a volcano on Sicily

Geosphere (lava and ash) materials and heat energy are released into the atmosphere.

5 List some Earth sphere interactions from your own daily activities.

Answers will vary. An example for transportation: The biosphere (humans) uses geosphere (metal) to manufacture vehicles. Engines use energy from the biosphere (oil) stored in the geosphere (underground) and add pollutants to the atmosphere.

6 Describe some global-scale interactions among Earth's spheres that are initiated by human action.

Answers will vary. Students may mention global warming, depletion of ozone, deforestation, or urban sprawl.

7 List and describe at least two new ideas you've had while considering Earth as a system.

Answers will vary. Look for evidence of students considering the entire Earth system rather than focusing on single spheres.

Teacher's Guide Chapter 1
Internet Investigation

How Do Interactions among Earth's Spheres Vary Regionally?   ▶ ES0108

1 For each location, tell where crops get their water. What Earth sphere interactions does each image indicate?

Crops in Argentina are watered by rain. This indicates that the hydrosphere is available at Earth's surface for the biosphere to thrive.

In Saudi Arabia, water is pumped from underground to water crops. This indicates the hydrosphere is scarce at Earth's surface and humans need to use energy to provide crops with water.

2 Which of the two images more closely represents how crops are watered in your region?

Answers will vary. For areas where irrigation is necessary, Saudi Arabia is the representative image. For areas where farmers depend on rain, the image of Argentine fields represents their region.

3 Could plants or animals survive in each of these places? Describe interactions between the geosphere and the biosphere for each image.

At Popocatépetl, the geosphere is destructive to the biosphere. Neither plants nor animals can live easily on this part of the geosphere. On the islands, the geosphere provides nutrients for the biosphere. Plants are abundant.

4 Describe interactions between humans and the geosphere illustrated by the images.

At mines, humans remove geosphere materials. At landfills, they return materials to the geosphere.

5 In your region, what materials do humans take from the geosphere? What materials are returned to it?

Answers will vary. Students may be aware of mines and landfills in your region.

6 Describe differences in interactions between the hydrosphere and the geosphere illustrated by these images.

At Capitol Reef, brief periods of contact with liquid rain wash sediments downhill. At Mount Rainier, frozen water grinds against rocks constantly. Differences in erosional processes result in varied landforms.

7 Describe interactions between the hydrosphere and the atmosphere near the equator. Compare these with interactions between the hydrosphere and the atmosphere near the poles.

Near the equator, the hydrosphere transfers large amounts of water vapor to the atmosphere. Near the poles, less water evaporates.

8 Describe interactions between the hydrosphere and atmosphere indicated by the evaporation rate for your location.

Answers will vary. Complete responses should include a general description of the evaporation rate in your region.

9 What is the Leaf Area Index for your own region in the image?

Answers will vary with your location.

10 Describe interactions among the biosphere, hydrosphere, and atmosphere indicated for your region by this image.

Answers will vary. For areas with abundant leaf coverage, the biosphere is actively transferring materials from the hydrosphere into the atmosphere. For areas with little leaf coverage, interactions are less active.

**Teacher's Guide Chapter 1
Internet Investigation**

1 Brainstorm with a couple of other students: Make a list of the kinds of information you would want to gather to help you predict where wildfires might break out.

Answers will vary. Students may list the dryness of vegetation, the length of time since it last rained, or the number of careless campers in a region

Students tables (below) will vary. Students use their own relative rating system to characterize fire potential. The average ratings (number 9) should be a number between 1 and 5. Ratings near 1 indicate a low fire potential; ratings close to 5 indicate a high risk of fire.

	Site A	Site B	Site C	Site D	Site E
2 Relative Greenness					
3 Departure from Avg.					
4 Live Plant Moisture					
5 Temperature					
6 Relative Humidity					
7 Wind Speed					
8 TOTAL					
9 AVERAGE RATING					

10 According to your ratings, which of the five sites has the highest risk of fire? Which site has the lowest risk?

Answers will vary depending on students' rating system. The national fire danger map shows that Site A has the highest fire danger and Site E has the lowest.

11 Describe how well or how poorly your fire potential ratings correlate with the national fire danger map.

Answers will vary depending on student ratings. With this simplified rating system, it is possible to calculate ratings very similar to the national fire danger map.

12 Recall the hypothesis you were testing: Rating and averaging six physical conditions of an area produces an accurate prediction of the area's fire potential. Based on the data you collected, should you accept or reject the hypothesis? Explain your answer.

Answers will vary. In most cases, data should support the hypothesis.

13 Of the six conditions you rated, which do you think are the most important predictors of fire? Which do you think are least important? Describe your reasoning.

Answers will vary. Students may suggest wind speed is most important because it blows sparks or that relative greeness is least important because plants are not yet mature.

14 How could you modify the rating system to give more weight to the most important predictors?

Answers will vary. Students may suggest assigning a mathematical weighting factor to each condition to indicate their relative importance in the final rating.

Teacher's Guide Chapter 2
Internet Investigation

How Might You Investigate Scientific Phenomena?

1 Describe some of the specific changes you observe in the animations.

Answers will vary. Students may mention seasonal changes of temperature.

2 Record the monthly estimates of vegetation levels for the last three months of 1986.

Mo.	Jan	Feb	Mar	Apr	May	Jun	Jul	Aug	Sep	Oct	Nov	Dec
UV Level	1000	2000	3000	4000	5000	6000	6000	5000	3000	2000	2000	1000
Veg. Index	0.1	0.2	0.3	0.4	0.5	0.5	0.5	0.4	0.3	0.3	0.3	0.2

3 Describe the trend of vegetation levels over the course of the year.

Vegetation steadily rises from January through May, then holds steady until July. After July, it decreases until September, then holds steady through November, at which point it drops again.

4 Describe the general relationship between vegetation and ultraviolet (UV) radiation levels, shown in the X-Y plot of the data.

In general, as UV levels increase, so do the vegetation levels.

5 Describe the trends in the levels of UV radiation over the course of the year.

UV level rises steadily from January through June. It holds steady through July, then decreases fairly steadily through December.

6 Based on the information presented here, would you accept or reject the hypothesis that the density of green vegetation is directly related to the amount of UV light that an area receives? Explain your answer.

Students should accept the hypothesis. There is a strong correlation between the two variables. The two graphs are very similar in shape.

Plans for Conducting a Scientific Investigation

I. Observing

Answers will vary

II. Ask Questions

Answers will vary

III. Form a Hypothesis

Answers will vary

IV. Design a research method

Answers will vary

V. Data Collection

Answers will vary

VI. Hypothesis Testing

Answers will vary

VII. Sharing your findings

Answers will vary

Teacher's Guide Chapter 2
Internet Investigation

How Do Map Projections Distort Earth's Surface?   ▶ ES0301

1 List at least three similarities and three differences among the maps.

Answers will vary. Similarities across the maps include location, scale, and color scheme. Differences include the type of data being mapped, the manner in which the data are symbolized, and the projections of the maps (which alters the shapes of the countries), as well as location, scale, and color scheme.

2 What happens to the surface of Earth when it is pulled off a globe and flattened?

When flattened, the surface has gaps. It does not form a continuous surface.

3 How do the two images compare?

The globe projection has gaps. The map projection is a continuous surface.

4 In which projection does Antarctica appear disproportionately large?

Antarctica is disproportionately large in the Miller projection.

5 In which projection does Asia appear disproportionately large?

Asia is disproportionately large in the Mercator projection.

6 Compare features and properties among the projections. Use the table to record your responses.

Answers may vary. Measurements are approximate.

7 Which projection minimizes distortion of South America, Africa, and areas near the equator?

The cylindrical projection minimizes distortion along the equator.

8 Which projection minimizes distortion of the United States and other temperate regions?

The conical projection minimizes distortion of the U.S. and temperate regions.

9 Which projection minimizes distortion of land in polar regions? Look at Antarctica or Greenland.

The planar projection minimizes distortion in polar regions.

Projection	Distance (km)	Area (km²)	Description of Shapes
Cylindrical	12,000	32,468,260	Africa appears narrow, U.S. appears wider than normal
Conical	6,800	34,134,728	Africa appears longer, U.S. shape close to actual
Planar	6.600	30,279,792	Africa appears much wider, U.S. appears wider than normal
Globe	6,826	30,279,792	True shapes

How Do Latitude and Longitude Coordinates Help Us See Patterns on Earth?

▶ **ES0303**

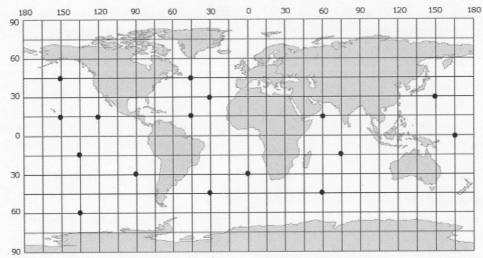

1 How does sea surface temperature change with latitude?

Accept all answers. (Any relationship is difficult to discern from the data in this form.)

2 How does sea surface temperature change with longitude?

Accept all answers. (No apparent relationship)

3 Which lines (longitude or latitude) run from north to south?

Longitude lines run from north to south.

4 Which lines (longitude or latitude) run from east to west?

Latitude lines run from east to west.

5 Which line is the reference line for longitudes?

The prime meridian is the reference line for longitudes.

6 Which line is the reference line for latitudes?

The equator is the reference line for latitudes.

7 Use colored pencils to trace these patterns onto the printed map. Describe the distribution of sea surface temperatures around the world.

Distinct bands of color are apparent. The warmest temperatures are found along the equator; the coldest areas are found at the poles.

8 How does sea surface temperature change along latitude lines?

Sea surface temp. remains constant along latitude lines.

9 How does sea surface temperature change along longitude lines?

Answers will vary. Sea surface temperature decreases from the equator to the poles.

10 What conclusion might you have drawn had you looked at global sea surface temperature along only one latitude line?

An analysis along only one latitude line might lead you to conclude that sea surface temperature remains constant on Earth.

11 How do sea surface temperature patterns change over the course of a year?

The bands of similar sea surface temperatures shift latitudes as seasons change.

How Are Landforms Represented on Flat Maps?   ▶ ES0307

1 Write a detailed description of the topography that you encounter during this flyby.

Answers will vary.

2 Compare the photo to the topographic map. Describe the pattern of the contour lines around features on the photo.

Answers will vary. Students may notice that contour lines surround hills and are more closely spaced when the topography is steep.

3 Which part of this land is the last to flood?

The flat top of the cliff is the last part to flood.

4 What is the elevation of the lines marked at A, B, and C?

The elevation at A is 6320 feet. The elevation at B is 6440 feet. The elevation at C is 6700 feet.

5 Describe the overall shape of the landscape.

This map shows a cliff with steep sides and a flat top.

6 What do closely spaced contour lines indicate about the shape of a feature? In other words, when the lines are close together, does the feature have gentle slopes or steep sides?

When contour lines are spaced closely on a map, the structure represented by the lines is very steep.

7 What is the pattern of the contour lines around a simple hill?

Contour lines around a hill form closed, nested loops that are round or oval.

8 Make a sketch of the pattern of the contour lines moving up the valley. Draw an arrow to indicate the direction in which water flows across the lines.

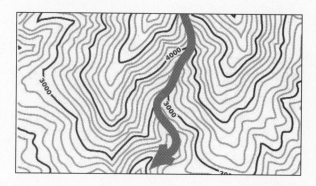

9 What landform feature does the model show, and what do hachures on contour lines indicate?

The model shows a volcano. The hachures show where the contour lines decrease in elevation. On the volcano, this occurs at the depression on its top.

10 Describe the structure inside the box on the map.

The structure is a valley descending from Mount Shasta.

11 Identify the features marked at A and B. Where is the elevation highest on this map? Where is it lowest?

A is a peak, B is a valley. The elevation is highest at the top of the northernmost peak. It is lowest at sea level (or at the base of the mountains).

12 Which of the landforms was easiest to recognize from its topographic map?

Answers will vary.

**Teacher's Guide Chapter 3
Internet Investigation**

How Can Getting Farther Away from Earth Help Us See It More Clearly?

1 Describe an advantage and a disadvantage of viewing Earth from farther away.

Answers will vary. Advantage: easier to see the "big picture" of how systems are interrelated. Disadvantages: higher cost and less detailed information.

2 Describe an advantage and a disadvantage of placing a satellite in a geostationary orbit.

Advantage: continuous observation of entire hemisphere. Disadvantages: requires multiple overlapping satellites for global coverage; poor coverage of polar regions.

3 Describe an advantage and a disadvantage of placing a satellite in a polar orbit.

Advantage: a single satellite can image the entire planet over time. Disadvantage: images from separate orbits must be mosaicked to reveal view of large area.

4 Why is active imaging necessary for measuring surface elevation?

Land at different elevations doesn't reflect light differently, or give off characteristic radiation. The detection system must actively bounce signals off the surface to measure elevation.

5 How do infrared wavelength detectors help weather forecasters?

Infrared radiation can be detected 24 hours a day, so forecasters have uninterrupted coverage of weather systems.

6 If a near-infrared image of vegetation were assigned to the red channel of a false color image, how would the vegetation appear?

Vegetation would appear red in the image, because it is very bright at near-infrared wavelengths.

7 Describe an example of a satellite-based study that would require the collection and analysis of temporal data.

Answers will vary. Any study that shows change over time would require temporal data.

8 Describe an example of a satellite-based study that would require the collection and analysis of multispectral data.

Answers will vary. Any study that requires the ability to discern different kinds of land surfaces would require multispectral data.

9 Choose one of the Earth observation satellites on this page and list the following: name, purpose, type of orbit (polar or geostationary), spectral bands, and ground resolution (the area each pixel represents).

Answers will vary. Example: Landsat 7, global environmental change, polar, 7 bands in visible and infrared, 30 meters per pixel.

How Can Getting Farther Away from Earth Help Us See It More Clearly?

Unit 1 Project Description

Use the results of this investigation plus Web-based research on Earth-observing satellites to design your own Earth observation satellite mission. Prepare a presentation that describes your satellite, its mission, and how it functions. Build a case for supporting your mission. The case you make for this new mission must be based on data and research. Present your findings in a format that has been approved by your teacher. Your project should include all the components listed below.

Unit Project Criteria

1. A description of the proposed satellite mission. (20%)
- Name the satellite
- Describe the purpose of the mission

2. An explanation of how the satellite functions. (30%)
- Describe the spectral characteristics (wavelength of radiation that it will use)
- Specify the data collection method (time series images of one area, or just one image of many places)
- Describe the type of orbit (polar or geostationary)

3. A case for supporting your mission. (35%)
- Provide examples of the kinds of questions you hope to answer using data the mission collects
- Establish a need for the unique data that your mission will collect
- Give a rationale for why this satellite should be built

4. An effective presentation of your conclusions. (15%)
- Communicate findings and ideas clearly and accurately
- Present appropriate data that support your conclusions
- Use appealing visuals and design

Earth System Interactions

Focus on the connections among Earth's spheres in your research and presentation. Think about how processes in the Earth system are revealed in satellite imagery. For example, consider:

- Remote sensing offers ideal opportunities for observing and recording Earth system interactions as they occur.
- Satellites track hurricanes and other storm systems, revealing atmospheric processes that alter the geosphere and destroy portions of the biosphere.
- Satellites monitor environmental changes such as deforestation—evidence of biosphere processes that affect the atmosphere by burning trees and affect the geosphere by increasing erosion.
- Satellites record fluctuations in sea surface height that indicate changes in hydrosphere circulation patterns, which change atmospheric patterns.

Teacher's Guide Unit 1
Internet Investigation

2 UNIT Earth's Matter

Face of the Earth™ ARC Science Simulations Copyright © 2001
www.arcscience.com

Explore how rocks change over millions of years

Read the clues in a rock to tell how it formed. Follow it through the rock cycle to see what it might become next.

Choose between paper or plastic

Design and conduct your own research project to decide which type of bag is best for the planet. Create an informational product to convince others of your choice.

Track an oil spill

Investigate how ocean currents and winds move oil spills. Learn how humans try to minimize the damage from these environmental accidents.

Unit 2
Earth's Matter

Earth Science

UNIT Earth's Matter

OTHER WEB RESOURCES

VISUALIZATIONS

Spark new ideas for investigations with images and animations, such as:

• Origin of the Solar System

• Earth cross-sections

• Rocks under a microscope

• 3-D models of molecules

DATA CENTERS

Extend your investigations with current and archived data and images, such as:

• Rocks and minerals

• Methods of keeping time

• Solar Power

• National Parks

EARTH SCIENCE NEWS

Relate your investigations to current events around world, such as:

• Environmental news

• Satellite launches

• Earth science discoveries

Unit 2
Earth's Matter

1 What evidence do you see of the dynamic nature of our planet?

Answers will vary. Students may mention swirling clouds, the African rift valley, or coastlines experiencing erosion.

2 Describe how the material is flowing.

The material appears to be rising and sinking as it flows.

3 What types of material do P waves pass through?

P waves pass through both solids and liquids.

4 What types of material do S waves pass through?

S waves pass through solids, but not through liquids.

5 Based on the pattern of the P and S waves, what type of material is this planet made of??

The entire planet is solid.

6 Observe the path taken by P and S waves in the model planet. Sketch the layers on your diagram and indicate if they are solid or liquid.

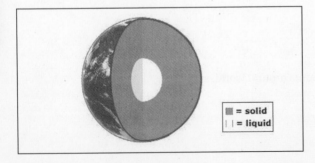

7 Observe the path taken by P and S waves in the model planet. Sketch the layers on your diagram and indicate if they are solid or liquid.

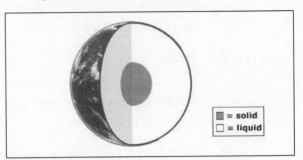

8 What happens to the size of the S wave shadow zone as the diameter of the liquid layer increases?

With increasing diameter of a liquid layer, the size of the S wave shadow zone becomes larger.

9 Observe the path taken by P and S waves on Earth. Sketch the layers on your diagram and indicate if they are solid or liquid.

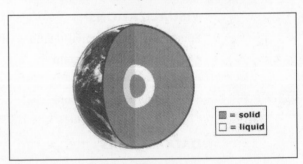

10 Where are the cool and hot regions of the mantle located?

Cool regions occur at the centers of large plates, while the warmest regions are found at the edges, especially along spreading ridges.

11 What does this tomographic model indicate about the underlying structure?

The subducting plate is moving faster than the surrounding rock materials. The plate sinks and cools as it moves under the adjacent plate.

Teacher's Guide Chapter 4
Internet Investigation

What Time Is It?

1 Which parts of Earth are experiencing day?

North America, South America, most of the Atlantic ocean, and about half of the Pacific ocean are experiencing day.

2 Which parts of Earth are experiencing night?

Europe, Asia, Africa, Australia, and about half of the Pacific Ocean are experiencing night.

3 In what direction does Earth rotate?

Earth rotates from west to east. Viewed from above the North Pole, the direction of Earth's rotation is counterclockwise.

4 From what direction does the sun appear to rise each day?

The sun appears to rise in the east.

5 In what direction does the sun appear to set each day?

The sun appears to set in the west.

6 When it is sunrise on the east coast of the United States, where on Earth is the sun setting?

The sun is setting along the eastern edge of Asia.

7 When it is sunrise on the west coast of Africa, approximately what "suntime" is it in India?

It is about noon in India.

8 Sketch the regions of night and day onto the map on your worksheet. Label the sunrise and the sunset boundaries. Circle the position of India.

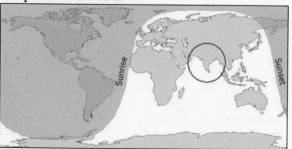

9 What happens to the shadow cast by an object as the sun appears to move across the sky?

Shadows cast by objects move in a clockwise direction in the northern hemisphere and counterclockwise in the southern hemisphere.

10 If Earth rotates 360 degrees every 24 hours, how many degrees does it rotate each hour?

In one hour the Earth rotates 15 degrees.

11 How many degrees of longitude separate Cairo and Calcutta, and how many time zones apart are they?

Sixty degrees of longitude separate the two cities. They are four time zones apart.

12 Is it earlier in the day in Cairo or Calcutta?

The hour is earlier in Cairo because it has not yet experienced sunrise.

13 When it is 6:00 A.M. in Calcutta, what time is it in Cairo?

It is 2:00 A.M. in Cairo.

14 How many hours apart are St. Louis, Missouri, U.S.A. and Kyoto, Japan?

Nine hours separate the two cities counting from St. Louis west to Japan; fifteen hours separate them counting from St. Louis east to Japan.

15 During what hours that you are awake can you call your friend so that it is not the middle of the night in Japan?

If you call your friend between 6:00 P.M. and 10:00 P.M. in St. Louis, then it will be between 9:00 A.M. and 1:00 P.M. in Kyoto.

**Teacher's Guide Chapter 4
Internet Investigation**

Earth Science

How Many Protons, Neutrons, and Electrons Are in Common Elements?

▶ ES0501

1 What elements are in these common items?

Diamonds, carbon; Balloons, helium; Rusting car, iron and oxygen; Nuggets, gold; Lighted sign, neon; Beverage can, aluminum.

2 List the three subatomic particles that make up atoms. Give the mass and charge of each one.

Proton: Mass = 1 atomic mass unit, +1 charge

Neutron: Mass = 1 atomic mass unit, neutral

Electron: Mass = 0, -1 charge.

3 Fill out the chart for these elements. Record the number of protons, neutrons, and electrons for *balanced atoms* of each element in the chart.

Name	Hydrogen	Helium	Lithium	Carbon	Nitrogen
Symbol	H	He	Li	C	N
Protons	1	2	3	6	7
Neutrons	0	2	4	6	7
Electrons	1	2	3	6	7
Weight (amu)	1	4	7	12	14

4 Which particle controls what element an atom is? Describe how you used the model to come up with your answer.

The number of protons in the model controls what element the atom is.

Answers will vary. Students may respond that when they added protons to the model, the name of the element changed.

5 What do you get when you change the number of neutrons in the nucleus?

Changing the number of neutrons makes an isotope of the atom displayed.

6 What controls the "weight" of an atom? Describe how you used the model to come up with your answer.

The number of protons plus the number of neutrons equals the weight.

Answers will vary. Students may respond that the weight always reflected the total number of protons plus neutrons they had in the model.

7 Try to cluster the electrons together or move them into another level. Describe the behavior of the model electrons.

The electrons repel one another. They move as far apart as possible. No more than two electrons are ever in the first level (K shell).

8 What do you get if the number of protons and electrons in your model is not equal?

An unbalanced atom.

9 Fill out the chart for these elements.

Name	Oxygen	Neon	Aluminum	Iron	Gold
Symbol	O	Ne	Al	Fe	Au
Protons	8	10	13	26	79
Neutrons	8	10	14	30	118
Electrons	8	10	13	26	79
Weight (amu)	16	20	27	56	197

**Teacher's Guide Chapter 5
Internet Investigation**

How Do Crystals Grow?

1 Write a description of how crystals grow in size.

Answers will vary. The ends or points of crystal structures appear to grow longer. New branches appear along the edges and extend outward.

2 What types of crystals exist, and where are they found?

Answers will vary. The list may include ice, snow, gemstones, geodes, quartz, evaporating saltwater or sugar solutions, metals, minerals, and rocks.

3 Describe similarities shared by all crystals as they grow.

Answers will vary. All crystals get larger over time, adding new material along points or edges at the outside of the crystal.

4 Describe ways in which the growth of each of the crystals is different.

Answers will vary. Crystals appear to grow in different shapes. Sometimes the shape is wider than it is long; sometimes cube-shapes form, and sometimes flat crystals form.

5 Based on their crystal lattice structures, predict the shape of galena and quartz crystals.

Galena crystals would have a cubic shape, while the quartz crystal would have a 6-sided (hexagonal) column shape.

6 Describe what happens when one growing crystal encounters another growing crystal.

When one crystal encounters another, the forward growth is halted, but growth can continue in other directions. Crystals may interlock, growing into available space around each other.

7 Describe the effect of temperature on the growth of these crystals from a solution.

Higher temperatures result in slower crystal growth.

8 Describe the effect of pressure on the growth of these crystals from a solution.

Higher pressures result in faster crystal growth.

**Teacher's Guide Chapter 5
Internet Investigation**

How Do Rocks Undergo Change? ▶ **ES0602**

Write a detailed description of the rock-forming processes represented by each set of arrows.

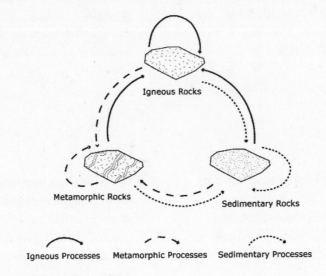

1 **Processes that result in igneous rocks**

Rocks are buried, subjected to great heat, and melted. Molten rock cools and crystallizes underground or erupts and crystallizes at the surface to form igneous rocks. The process of igneous crystallization is analogous to freezing liquid water to make ice.

2 **Processes that result in sedimentary rocks**

To make a sedimentary rock, the original rock experiences uplift, weathering, erosion, transport, deposition, compaction, and cementation.

3 **Processes that result in metamorphic rocks**

An increase in temperature or pressure or both changes rocks into metamorphic rocks. If temperatures continue to increase, metamorphic rocks may melt, then recrystallize as igneous rocks.

How Do Igneous Rocks Form?

1 Describe the four igneous rocks shown in the images.

Answers will vary.

Sample 1

Reddish color

Large visible, interlocking crystals

Sample 2

Black color

Looks like glass

Sample 3

Red color with white spots

Fine-grained background

Sample 4

Dark color

No visible crystals

2 Identify the texture of each of these igneous rocks as coarse-grained, fine-grained, or porphyritic. Record your answers in the table.

3 Identify the cooling rate (fast, slow, or two-staged) and cooling environment (magma chamber, eruption from volcano, or deep cooling followed by eruption) of the rock in each image. Record your answers in the table.

Rock	Texture	Cooling Rate and Environment
Sample 1	Coarse	Slow cooling rate; formed in magma chamber
Sample 2	Fine	Fast cooling; erupted from volcano
Sample 3	Fine	Fast cooling; erupted from volcano
Sample 4	Porphyritic	Two-staged–slow, then fast. Deep cooling followed by eruption
Sample 5	Porphyritic	Two-staged–slow, then fast. Deep cooling followed by eruption
Sample 6	Coarse	Slow cooling rate; formed in magma chamber

Teacher's Guide Chapter 6
Internet Investigation

What Kind of Rock Is This?

For each rock sample:

1 Record the sample number on your table.

2 Describe the texture and color or composition of your sample.

3 Choose the statement that best describes your sample. Click the statement and continue choosing the best statement to describe your sample.

4 Compare features of your sample with the image to make sure your identification is reasonable. Record the rock type and rock name in the table.

Sample Number	Description	Rock Type	Rock Name
	Answers will vary.		

1 Predict what you think happens when an oil spill like this one occurs.

Accept all responses.

2 As you explore, write down three observations about oil spills.

Accept all observations.

3 What was the average rate at which the oil spread?

8.39 miles per day.

4 List at least two factors that might affect the way the oil moves.

Wind speed, wind direction, currents in the ocean, and the type of oil involved in the spill can all affect the way the oil moves.

5 Predict what will happen to the oil if winds are blowing from the south at 15 knots per hour.

Responses will vary but may include the idea that the oil will change direction and flow to the north.

6 Describe how the oil moves now.

The combined effects of the wind and the current cause oil to move to the northeast. The addition of winds from the south causes the oil to now hit the shoreline.

7 How does what you see compare with what you predicted would happen?

Responses will vary. Accept all answers.

8 How does the increase in wind speed affect the way the oil moves?

The higher the wind speed, the faster the oil reaches the shore. Oil propelled by high winds does not spread out as far along the shoreline

9 Use different colors or symbols to mark the areas of impact during each wind condition on the map below.

10 What areas of the shoreline will be affected by the spill?

Seabird nesting and shellfish beds to the west of Sachem Head are now affected.

11 Sketch the areas of impact on the map.

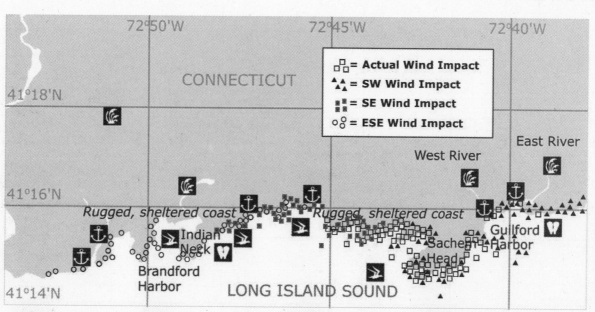

 Anchorage   Shellfish beds   Marshland   Seabird nesting on rocks & islets

Teacher's Guide Chapter 7
Internet Investigation

Why Is This Place Protected?   ▶ ES0705

1 What types of things make a place worth protecting? List at least three things that you think are important.

Accept all answers.

3 List five renewable and five nonrenewable resources.

Forests, crops, water, air, and soil are renewable resources. Coal, petroleum, sand, salt, sulfur, and many other minerals are nonrenewable.

2 Record your observations in the table.

Answers will vary. As an example, in the table below, a set of observations is recorded for Yellowstone National Park. Look for similar responses from students.

4 Create a brochure that highlights the unique features of this place. Clearly explain why this place should be protected.

Answers will vary. Look for completeness.

Protected Place	State	Description of unique features	Major attraction (What do people do there?)	Natural Resources	What might happen if this place were not protected?
1. Yellowstone National Park	WY	geysers	Observe the thousands of hot springs as well as the wildlife and vegetation	geothermal energy; bears, fish, and other wildlife; mineral deposits	Geysers might be destoyed; Wildlife might be hunted to the point of extinction
2.					
3.					
4.					

What Environmental Changes Can We See with Satellites?   ▶ ES0707

1 As you explore the images, write at least three questions that come to mind about the growth of Las Vegas.

Answers will vary. Accept all questions.

2 Compare the color of vegetation in each image.

Vegetation is dark gray in the Corona image and maroon to bright red in the Landsat image.

3 What new features do you see each year?

From 1964 to 1972 the number of paved roads increased. In 1986 there is a new lake and a shopping mall that were not present in the 1972 image. In 1992 an additional golf course, lakes, and many more housing developments are present.

4 List at least two criteria you will use to determine if an area has experienced urban development.

Accept reasonable criteria such as the presence of roads and grids of vegetation.

5 How much of the area of a square should contain developed land if you are to include its count in the overall measurements?

At least fifty percent or more of the square should be developed to include it in the count.

6 Record your area measurements in the table.

Answers may vary.

Year	Area in Square Miles	Population
1964	32	127,000
1972	48	273,000
1986	95	608,000
1992	169	863,000
1997	—	1,124,000

7 How has the area of the city changed through time?

The city has spread out in all directions, but the largest increases in area have been in the northwest and southeast.

8 What factors might explain why some parts of the city are growing at faster rates than others?

Factors such as population density, the availability of water, the topography of the area, and the cost of land in different parts of the city can account for the environmental changes.

9 Describe the shape of your graphs and explain how the growth rates change through time.

Both graphs show exponential growth. When growth rates are exponential, their graphs curve steeply upward, indicating that growth increases faster and faster.

10 What would happen to the shape of the graphs and the growth rates if the data from only two years were plotted?

The only graph that can be generated using two points is a straight line. Thus, the growth rate would appear to be linear and the rate of change assumed to be constant in each graph.

11 When it comes to urban growth, what environmental changes can satellite images show?

The extent of development, the rate of development, and the specific types of changes in land use can be observed with satellite images taken over time.

12 List at least three other environmental changes that satellites can help us monitor.

These might include human-induced changes such as deforestation and water depletion, or changes brought about by natural hazards such as those resulting from volcanoes and earthquakes.

Teacher's Guide Chapter 7 Internet Investigation

Paper or Plastic–Which Type of Bag Is Better for the Environment?

▶ **ESU201**

1 In a single sentence, summarize the goal of this step. Write down a few questions or ideas that will help you start brainstorming a list of issues to explore in the paper-versus-plastic debate.

The goal is to generate a list of issues related to the paper-versus-plastic question.

Lists of issues will vary. Students may consider the availability of the raw materials used to produce each bag or the cost to consumers.

2 In a single sentence, summarize the goal of this step. Record your ideas for research methods based on your interests and the resources available to you.

The goal of this step is to gather information on a variety of possible topics relating to the problem. Ideas for research methods will vary with interests and availability of resources. Students may consider doing strength tests on bags or comparing costs.

3 In a single sentence, summarize the goal of this step. Describe at least one issue in the paper-versus-plastic debate from both sides of the argument.

The goal of this step is to list positive and negative issues associated with each side of the argument. On the issue of efficiency, for example, paper bags hold more, but more plastic bags can be carried at a time.

4 In a single sentence, summarize the goal of this step. List three factors you currently believe would provide convincing evidence of one bag's superiority over the other. (Don't support one type of bag or the other at this point; simply list factors you think could convince you that one is better than the other.)

The goal of this step is to organize a list of factors that form the basis of a convincing argument in favor of one position. Lists will vary. Factors that might provide convincing evidence include lower production costs, greater strength of material, or speed of decomposing once thrown away.

5 In a single sentence, summarize the goal of this step. List at least two final presentation ideas that appeal to you.

The goal of this step is to prepare a formal presentation in favor of one type of bag over the other. Lists of presentation ideas will vary. Students list written reports, creation of a brochure or a poster, or a skit with grocery customers giving reasons for the type of bags they choose.

6 Based on what you currently know about paper and plastic bags, describe at least one reason that humans should make a research-based decision in choosing one type of bag over the other.

Answers will vary. The depletion of non-renewable resources in the closed Earth system is one reason to make a research-based decision for one type of bag.

Paper or Plastic–Which Type of Bag Is Better for the Environment?

Unit 2 Project Description

How does a choice as simple as deciding which type of bag to use at the grocery store affect the larger Earth system? Using Web resources and other research methods, consider the paper-versus-plastic question from many perspectives and make a decision based on your findings. Select one or more factors relating to the decision of which type of bag to use, then gather information to present a balanced argument in favor of one of the options. Present your findings in a format approved by your teacher. Your project should include all components listed below.

Unit Project Criteria

1. **A comprehensive list of issues related to the question (10%)**
   - Develop an extensive list of issues related to the manufacture, use, and disposal of paper and plastic bags.
   - Include ideas that come from a variety of different categories.

2. **Evidence of credible research (30%)**
   - Demonstrate that a number of different sources were used for research.
   - Show that a variety of types of resources were consulted.
   - Present evidence of research in the form of useful notes.
   - Make appropriate references to the sources you consulted.

3. **A balanced argument (20%)**
   - Provide information on both types of bags in a balanced manner.
   - Identify potential biases in your sources of information.

4. **A credible case for your position (20%)**
   - Take a stand on one side of the issue.
   - Defend your position logically and support your arguments with citations or data.

5. **A presentation of your research findings (20%)**
   - Communicate findings and ideas clearly and accurately.
   - Present appropriate data.

Earth System Interactions

Focus on the connections among Earth's spheres in your research and presentation. Think about the implications of using paper or plastic bags for our planet as a whole. For example, consider:
- the geosphere and biosphere, which are the sources of raw materials for bags
- the atmosphere, which absorbs by-products from the production of each type of bag
- the hydrosphere, which absorbs chemical by-products of bag production

3 UNIT Dynamic Earth

NASA

Check out active volcanoes

Access volcano-cams to look for current activity on volcanoes around the world.

Investigate earthquakes in Los Angeles

Analyze ground motion around Los Angeles to pinpoint the location of a blind thrust fault.

Explore mountain belts

Check out some of Earth's most prominent mountain chains. Examine photos of geological features from each range to figure out how the mountains formed.

Unit 3
Dynamic Earth

UNIT 3 Dynamic Earth

OTHER WEB RESOURCES

VISUALIZATIONS

Spark new ideas for investigations with images and animations, such as:

• Motion of the plates

• Different types of volcanoes

• Earthquake videos

• Mountains forming

DATA CENTERS

Extend your investigations with current and archived data and images, such as:

• Live volcano-cams

• Earthquake seismic data

• Safety tips

• Volcanoes on other worlds

EARTH SCIENCE NEWS

Relate your investigations to current events around world, such as:

• Active volcanic eruptions

• Earthquakes around the world

• New discoveries about tectonics

Unit 3
Dynamic Earth

What Is Earth's Crust Like?   ▶ ES0801

1 Hypothesize about why volcanoes form linear patterns across the globe. What does this pattern tell you about Earth's outer shell, or crust?

Answers will vary. Heat from Earth's interior escapes through volcanoes, so these lines might be weak points within the crust that allow the heat to break through.

2 How deep are the deepest earthquakes?

600-700 km

3 What does the lack of earthquakes below the depth you reported in question 2 suggest about the rocks at that depth?

Answers will vary. Below the level of the deepest earthquakes, temperatures are high enough that rocks under stress do not break and generate earthquakes.

4 In what direction was pressure applied to create the Himalaya Mountains?

Approximately north-south.

5 How does the orientation (direction) of folded mountain belts compare with nearby zones of volcanoes and earthquakes?

Folded mountains are roughly parallel to nearby zones of volcanoes and earthquakes.

6 What do chains of folded mountains tell you about Earth's crust?

Earth's crust absorbs pressure applied in a direction perpendicular to zones of earthquakes and volcanoes. The crust acts as a rug being pushed into wrinkles.

7 From your examination of the data, describe your mental model of Earth's crust.

Answers will vary. Earth's crust is solid in some areas, but it is crossed by zones where heat escapes and earthquakes occur. Horizontal pressure causes the crust to "wrinkle" in some areas.

8 On the world map, outline the areas where Earth's internal heat escapes and solid rocks move against other solid rocks. Show folded mountain belts and draw arrows to indicate the direction in which pressure was applied to form them.

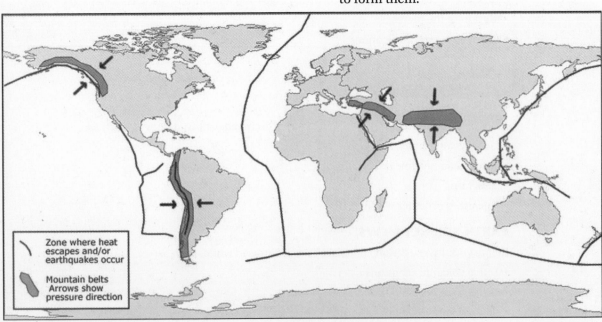

Zone where heat escapes and/or earthquakes occur

Mountain belts Arrows show pressure direction

How Old Is the Atlantic Ocean?   ► ES0802

1 In addition to their shapes, what other data support the hypothesis that Africa and South America were once joined together?

Matching types and ages of fossils, rocks, and mountain ranges on South America and Africa support the hypothesis that they were once joined.

2 Along the profile line, where is the ocean generally the deepest? ...the shallowest?

The Atlantic Ocean is deepest just off the continental shelves, near the continents, and shallow in the center, along the Mid-Atlantic Ridge.

3 What areas of the Atlantic seafloor have the youngest rocks? ...the oldest?

The youngest rocks are along the Mid-Atlantic Ridge. The oldest rocks are near the continents.

4 How old are the very oldest rocks on the Atlantic seafloor? Where are they?

The oldest rocks are about 180 million years old. They are off the coasts of N. America and Africa.

5 Based on the age of the oldest rocks between South America and Africa, when did the two continents split?

The continents split about 118 million years ago.

6 Based on ages of the oldest rocks in the North and South Atlantic, describe how and when the Atlantic Ocean formed, and how its shape has changed through time.

The Atlantic Ocean started as a linear sea, cutting across the joined continents from the north about 180 million years ago. The ocean extended only as far south as the current location of N. America and N. Africa, until about 118 million years ago when S. America and Africa split. The ocean was initially linear, with roughly the same shape as the mid-Atlantic ridge. The ocean has been getting wider for 118 million years.

7 Measure the distance between South America and Africa.

About 5100 km

8 Calculate the average yearly widening of the Atlantic Ocean in cm per year. Show your work.

Answers may vary slightly.

5100 km / 118 million years = 4.3 cm per year

9 Where else on Earth do mid-ocean ridges exist?

The Indian Ocean and Pacific Ocean each contain mid-ocean ridges.

10 What can you infer about continents on opposite sides of the same mid-ocean ridge?

They may have fit together at one time.

11 What evidence would you look for if you wanted to show that two landmasses were once connected?

Answers may vary but should include items such as matching rock types, fossils, or mountain ranges on currently separated continents.

Copyright © McDougal Littell Inc.

Teacher's Guide Chapter 8
Internet Investigation

Earth Science   *Internet Investigations Guide*   **33**

How Fast Do Plates Move?   ▶ ES0810

1 Which island do you think is youngest? …oldest? Explain your reasoning.

Answers will vary. Active volcanoes on Hawaii and inactive volcanoes on islands to the northwest indicate Hawaii is youngest and Nihoa is oldest.

2 What pattern do you see between age of rocks and location?

Age of rocks increases with distance northwest of the island of Hawaii.

3 How do you think the chain of islands might have formed? Explain your idea.

Accept any justified answer. The islands formed as the Pacific plate moved over a hotspot. The plate has moved to the northwest over the stationary hotspot. New volcanoes form over the hotspot as old volcanoes are carried northwestward by the plate.

4 How can you use the map information to calculate the rate (speed) of plate motion over the hotspot? Describe the information you need and how you'll use it.

Information required is the distance the plate has moved and time over which the movement has occurred. Distance / Time = Rate

5 Calculate the average rate of the Pacific plate's motion over the past 5.1 million years. Express your answer in cm/year.

Answers may vary.

600 km / 5.1 million years ≈ 10 cm/yr

6 Calculate the average rate of plate motion for 0 – 40 million years ago.

Anxwers may vary.

3500 km / 40 million yr ≈ 9 cm/yr

7 Calculate the average rate of plate motion for 40 – 60 million years ago.

Anxwers may vary.

2300 km / 20 million yr ≈ 11 cm/year

8 What does the bend in the chain of seamounts indicate?

A change in the direction of plate motion

9 What global tectonic event might have been responsible for changing the direction of the Pacific plate's motion? (What major tectonic event occurred around 40 million years ago?)

The collision of India into Asia to form the Himalayas.

10 What evidence is there of a hotspot currently under Yellowstone National Park?

Geysers indicate a heat source below Yellowstone.

11 In what direction has the North American plate moved over the hotspot under Yellowstone? Describe your reasoning.

To the southwest. The trail of smooth topography gets younger to the northeast.

12 Calculate the average rate of the North American plate's motion over Yellowstone.

Answers may vary.

300 km / 10 million yr ≈ 3 cm/yr

How Are Volcanoes Related to Plate Tectonics? ▶ ES0901

1 Were your predictions for plate boundary locations close? What clues did you use to make your most accurate predictions?

Answers will vary. Topography of the seafloor is the best predictor of plate boundary locations.

2 Sketch the location of volcanoes and plate boundaries onto your map.

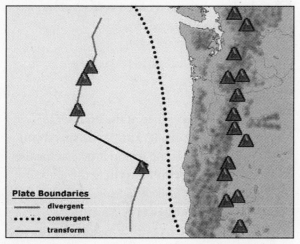

Plate Boundaries
- ———— divergent
- • • • • • convergent
- ———— transform

3 Why do you think volcanism on the seafloor occurs directly along a plate boundary, but the Cascade volcanoes occur along a line at some distance from the plate boundary?

Answers will vary. Volcanism on the seafloor is related to a divergent boundary. Magma from directly below the boundary fills in the void as plates move apart. The Cascade volcanoes are related to subduction. The subducted plate travels downward and east before it reaches a point hot enough to generate magma. This magma rises upward to create volcanoes on the surface.

4 Sketch the cross section of this region onto your worksheet.

5 Where does the magma for volcanism at the rift zone come from?

Magma for rift zone volcanism comes from melted mantle material.

6 What material is melted to provide magma to the Cascade volcanoes?

Subducted crustal material is melted.

7 Sketch a cross section showing the subsurface geology you would expect to see along A-B.

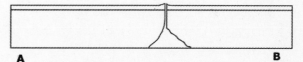

8 Where does the magma for these volcanoes come from?

Melted mantle material is the source of magma.

9 Sketch a cross section showing the subsurface geology you would expect to see along A-B.

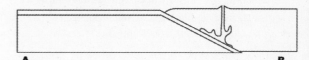

10 What type of eruptions would you expect these volcanoes to have? Cite evidence for your answer.

These volcanoes would erupt explosively. Like the Cascades, these are subduction-related volcanoes, so they would exhibit a similar type of eruption.

Teacher's Guide Chapter 9
Internet Investigation

Pinatubo

How Fast Do Gases from Volcanic Eruptions Travel?   ▶ ES0906

1 Describe how you think this eruption might have looked from space.

Answers will vary. When viewed from space, the eruption appeared as a dark circular area.

2 How do you think people in Manila were feeling that day? Describe what you might have been doing if you were there.

Answers will vary. Accept any reasonable answer.

3 What happens to the *size* of the sulfur dioxide (SO_2) cloud over time?

The size of the SO_2 cloud increases over time.

4 What happens to the concentration of SO_2 over time?

The concentration of SO_2 decreases over time.

5 Does the lack of a defined SO_2 cloud on June 30 mean that the SO_2 was no longer in the atmosphere? Explain your answer.

No. The concentration of SO_2 had fallen below detectable levels because it was spread out over a larger area. The total amount of SO_2 had not changed appreciably.

6 What information would you need to calculate the speed of the SO_2 cloud?

The distance it had traveled over a given time.

7 Use the distance the SO_2 gas traveled and the time since the eruption to calculate its speed in km/hr. Show your work.

5,000 km/5 days = 1000 km/day

1000 km/day / 24 hours/day ≈ 40 km/hour

8 How is the movement of the aerosol cloud similar to the sulfur dioxide cloud? How is it different?

Answers will vary. Both clouds moved in the same direction, but the SO_2 appeared to spread out over a larger area and the aerosol cloud appeared to get smaller.

9 Calculate the speed of the aerosol cloud in km/hr. Show your work.

5,000 km / 5 days = 1000 km/day

1000 km/day / 24 hours/day ≈ 40 km/hour

10 Earth's circumference at the equator is approximately 40,000 km. At the speed you calculated, how long would it take aerosols from Mount Pinatubo to circle the globe?

40,000 km / 40 km/hour = 1000 hours

1000 hr/day / 24 hr/day ≈ 40 days

11 What differences do you notice between the two photos? Which photo do you think shows evidence of aerosols in the upper atmosphere?

Answers will vary. The photo on the right shows gray-colored layers extending across the image. These might be small ash particles.

12 Describe one way you might test the idea that Mount Pinatubo's eruption was responsible for this instance of global cooling. Explain your answer.

Answers will vary. One test would be to examine climate records for months following other large volcanic eruptions.

Teacher's Guide Chapter 9
Internet Investigation

Is It Safe to Live Near a Volcano? ▶ ES0907

1 Based on this map, how much of the area surrounding Mount St. Helens would you evacuate? Explain your reasoning.

Answers will vary. Students may choose a large area to ensure everyone's safety, or a small area because they don't want to invade private lives.

2 On your chart, describe the impact in each zone. Give a descriptive name to each zone to indicate the cause of the destruction.

Zone #	Description of Impact	Name for zone
1	Answers will vary. Light colored rocks from the volcano flowed downhill from the main crater.	Volcanic rock flow
2	Huge rocks and other debris flowed downhill.	Debris flow
3	Water carrying huge amounts of volcanic ash flooded existing rivers, carrying trees downstream and depositing them on riverbanks.	Mud flow
4	Entire forests of mature trees blown over in one direction, directly away from the crater.	Blast zone
5	Trees are still standing, but have lost their needles, which perhaps were killed by heat.	Scorched zone

3 Do you think a Cascade volcano might erupt in your lifetime? Explain.

Accept any justified answer. Students might respond "No" because one just erupted in 1980.

4 Based on the graph, which Cascade volcano would you predict may erupt next? Describe your reasoning.

Accept any justified answer. Students may predict that Mount St. Helens would erupt again because it has a long history of frequent eruptions.

5 What steps would need to be taken to save lives if Mount Rainier were to enter an active eruptive phase?

Answers will vary. You would need to notify people of the danger and get them out of the area.

6 Draw a line around the area you would recommend for evacuation onto your map.

Answers will vary. Students should enclose an area that extends beyond areas covered by historical mudflows.

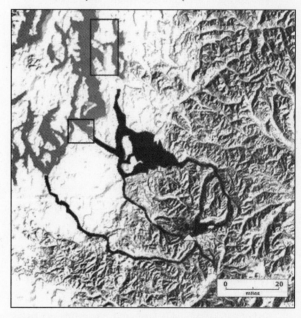

7 Outline a plan for safely and efficiently evacuating residents from the area you identified.

Answers will vary. Complete answers should provide a justification of the evacuation rules they establish.

Teacher's Guide Chapter 9 Internet Investigation

1 Predict where most earthquakes occur.

Answers will vary. Students may suggest that they occur along coasts.

2 Describe how the pattern of earthquake locations compares to the location of plate boundaries.

Most earthquakes occur along plate boundaries.

3 Which depth range has the most earthquakes? Which depth range has the fewest earthquakes?

Most earthquakes are shallow. Deep-focus earthquakes are least common.

4 At what type of plate boundary do you find most deep-focus earthquakes?

Deep-focus earthquakes are found at convergent plate boundaries or subduction zones.

5 Explain why deep-focus earthquakes occur at this type of plate boundary but not at others.

The contact area between adjacent plates extends to greater depths in subduction zones than at other boundaries.

6 Which magnitude range has the greatest number of earthquakes? Which has the fewest?

Earthquakes in the magnitude 1-3 are most abundant. Earthquakes in the magnitude range 7-9 are least common.

7 Which type of plate boundary experiences the most earthquakes with large magnitudes? Hypothesize about why that type of boundary experiences larger earthquakes than the others.

Convergent plate boundaries experience the most large-magnitude earthquakes. Student hypotheses will vary: Pressure between plates at convergent plate boundaries builds to higher levels before rocks slip, releasing larger amounts of stored energy per earthquake.

8 Give three examples of damage that can occur as a result of earthquakes.

Answers will vary. Students may mention fallen buildings, broken glassware, fires, damaged roads, and broken bridges.

9 Which cities are susceptible to earthquake activity? Describe the factors you considered in order to arrive at your answer.

Cities that are most prone to earthquakes are Los Angeles, Seattle, Memphis, Tokyo, and Manila. All these cities are near plate boundaries and/or have a history of earthquakes.

1 What do you notice about the time interval between the arrival of P and S waves at the three different seismograph stations? What causes these differences?

The difference between P and S wave arrival time is larger for stations farther from the epicenter. This is due to the difference in velocity between P and S waves.

2 For each location, record the distance to the epicenter.

Victorville   133 km to epicenter

Lancaster   71 km to epicenter

Los Angeles   42 km to epicenter

3 For each location, draw circles corresponding to the distances you recorded in question 2. Sketch the circles on your map and mark the epicenter location.

Epicenter is at 34° 15' N, 118°7.5' W

4 Describe three different examples of damage that occurred as a result of the Northridge earthquake.

Answers will vary. Answers might include fallen buildings, fires, cracked roads, and broken bridges.

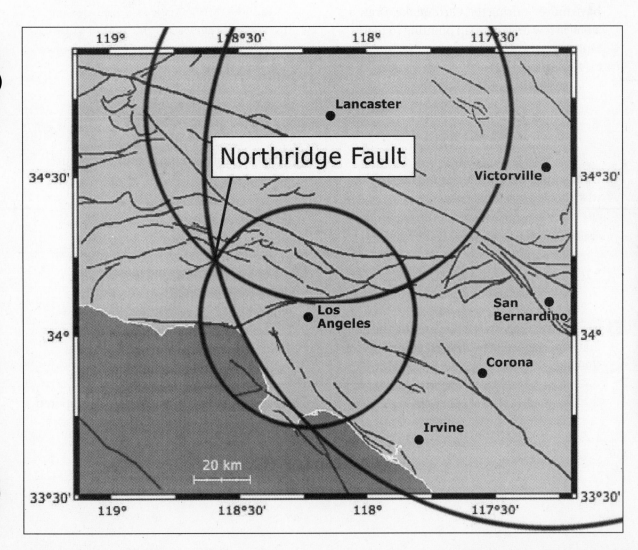

Copyright © McDougal Littell Inc.

Which Fault Moved in the Northridge Earthquake?

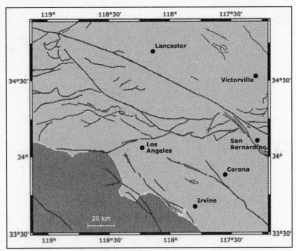

1 Make a prediction about which fault might have moved during this earthquake. Draw a small arrow on your map pointing to the fault.

Answers will vary. Students should choose a fault somewhere near the epicenter.

2 On the map, draw a line around the area that was affected by aftershocks.

Aftershocks were concentrated in an oval-shaped area near the Northridge epicenter.

3 In another color, draw a line around the area that experienced the strongest ground shaking.

The strongest ground shaking was concentrated in an oval-shaped area near the Northridge epicenter.

4 In a third color or pattern, draw a line around the area that had the highest velocity of ground motion.

Aftershocks were concentrated in an oval-shaped area near the Northridge epicenter.

5 Which fault was responsible for the Northridge earthquake? Highlight the fault on your map. What evidence did you use to arrive at your conclusion?

Northridge thrust fault, just north of the epicenter. (See map for ES1003 on page 39 for marked location of the Northridge thrust fault.) This fault is closest to the area of greatest ground motion, velocity, and location of aftershocks.

6 At what depth did the Northridge occur?

Around 18 km

7 At what depth range did most of the aftershocks occur?

0-10 km

8 Based on location of aftershocks compared to the Northridge epicenter, in what direction is the fault inclined?

Fault is inclined to the southwest.

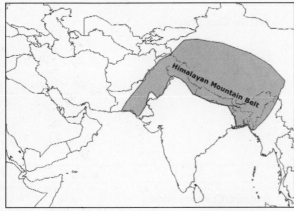

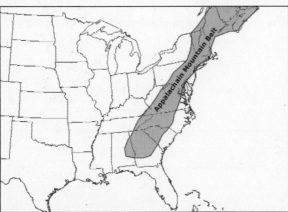

1 On your map of South America, sketch in the location of the Andes Mountains, and label the chain's length and width.

Answers will vary. They are about 6000 kilometers (km) long and 190-750 km wide.

2 What type of plate boundary occurs west of the Andes? Sketch and label it on your map.

Ocean-continent convergent boundary

3 On the map of Asia, sketch in the location of the Himalayan Mountain belt. Record its length and width.

Answers will vary; 2400 km wide and 3600 km long.

4 What type of plate boundary occurs near the Himalayas? Sketch it onto your map.

Continent-continent convergent boundary

5 On the map of North America, sketch in the location of the Appalachian Mountain belt, and record its length and width.

Answers will vary; 250-400 km wide, 900-1000 km long.

6 How do you think these mountains formed? What do you notice about the Appalachian Mountain belt and current plate boundaries?

The Appalachian Mountains probably formed near a convergent plate boundary. They are not located along a modern plate boundary.

7 Describe the geologic features and their ages for each mountain belt.

Andes Mountains: Stratovolcano and folds. Both features are recent. Himalayan Mountain belt: Folds and thrusts. Both features are recent. Appalachian Mountain belt: Plutonic rocks, approximately 500 million years old; fold and thrust faults, approximately 300 million years.

8 Compare the Andes, Himalayan, and Appalachian Mountain belts. Describe features that are common to all three.

The Andes are a long narrow mountain belt located along a subduction zone. The Himalayas are wide, with high mountains, along a convergent plate boundary. The Appalachian Mountain belt is short and narrow compared to the others. All three belts are marked by folds and thrust faults.

9 Hypothesize about why these mountain belts of the same age are separated by the Atlantic Ocean.

Divergent plate motion separated a mountain belt that originally formed near a convergent boundary.

Teacher's Guide Chapter 11
Internet Investigation

How Do Rocks Respond to Stress?

1 Which image shows an example of brittle deformation? Which image shows an example of ductile deformation? Cite your evidence.

The image of the fold shows an example of ductile deformation. Layers are continuous and have not been fragmented. The other image shows an example of brittle deformation. The rock has been fragmented and layers have been broken and disrupted.

2 Sketch the resulting structure for each animation. On your sketch, indicate the stress directions and conditions affecting deformation.

3 Identify the type of structure, type of deformation, and stress direction for each image.

A. Fold, Ductile, Compression

B. Fault, Brittle, Compression

C. Shear Zone, Ductile, Shear

D. Fault, Brittle, Tension

E. Boudin, Ductile, Tension

Brittle/Tension

Low temperature and pressure
Arrows pointed out

Brittle/Compression

Low temperature and pressure
Arrows pointed in

Brittle/Shear

Low temperature and pressure
Arrows side-by-side

Ductile/Compression

High temperature and pressure
Arrows pointed in

Ductile/Tension

High temperature and pressure
Arrows pointed out

Ductile/Shear

High temperature and pressure
Arrows side-by-side

Teacher's Guide Chapter 11 Internet Investigation

1 Sketch the fault including layers A and B. Label the fault plane, hanging wall, and footwall. Name the type of fault, and draw arrows to indicate the direction of stress.

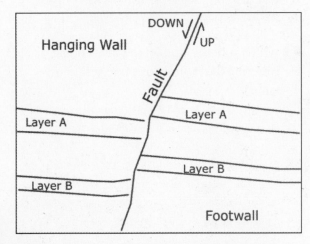

2 Sketch the fault including layers A and B. Label the fault plane, hanging wall, and footwall. Name the type of fault, and draw arrows to indicate the direction of stress.

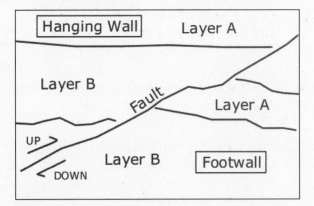

3 Describe how you deduced the direction of stress that formed this feature.

Layers in the hanging wall have moved downward relative to the footwall.

4 What can you infer about the forces that built the mountains of the Basin and Range province?

Mountains in the Basin and Range province are all produced by tensional forces.

5 Describe how you deduced the type of stress that formed this feature.

Layers in the hanging wall have moved upward relative to the footwall.

6 What type of stress was responsible for building the northern Rocky Mountains?

Compression.

7 What type of faulting is indicated by the cross section of the Rio Grande rift?

Normal

8 What type of stress formed the mountains near Albuquerque, New Mexico?

Tension

9 What type of force was responsible for creating the folds and faults in the Appalachian Mountains?

Compression

10 What tectonic process might be responsible for applying this force over an area as large as the Appalachian Mountains?

Continent-continent convergence.

Teacher's Guide Chapter 11
Internet Investigation

How Can One Volcano Change the World?   ▶ ESU301

Pinatubo

1 What is the plate tectonic setting of the Philippine Islands? Describe the geologic processes and features you would expect to find there.

The Philippines are near a subduction zone; oceanic lithosphere is subducted beneath the islands. Explosive volcanoes and earthquakes that range from shallow to deep occur there.

2 List some of the pre-eruption interactions you identified for each of Earth's spheres.

Answers will vary. Vegetation covers the geosphere, moving hydrosphere from the soil to the atmosphere through transpiration. Hydrosphere rains out of the atmosphere and runs down sides of mountain to the ocean. Geosphere and hydrosphere meet at the beach. Humans grow crops (biosphere) for food. Mountains (geosphere) force atmosphere to higher elevations, causing water vapor to condense into clouds.

3 List a few more descriptions of pre-eruption interactions so that you have considered many aspects of the Earth system.

Answers will vary. Humans use geosphere materials to build ships to import or export products. Humans catch fish and grow crops (biosphere) to feed themselves and their livestock.

4 How did the eruption change interactions among Earth's spheres? Describe how processes you listed in #2 and #3 changed as a result of the eruption.

Answers will vary. The geosphere killed and buried vegetation and people (biosphere). Heat energy plus ash (geosphere) were blasted into the atmosphere, changing its temperature and composition. Ash was suspended by water (hydrosphere) and transported to another location on the geosphere. Humans used geosphere materials to build dikes to protect the biosphere from lahars. Biosphere materials on the forested slopes of the mountain were buried; they eventually become part of the geosphere.

5 Describe post-eruption interactions illustrated by the images.

The shape of the geosphere changed: the mountaintop became a crater, and thick accumulations of ash were deposited on the surface. Rivers were filled in with geosphere material. Buildings were buried.

6 Which spheres interact to form lahars? Describe the interactions that result in lahars.

Geosphere, atmosphere, and hydrosphere interact to form lahars. An eruption from the geosphere placed loose ash on the surface. Atmospheric circulation patterns brought abundant rain to the region. The hydrosphere lubricated the geosphere materials and they flowed downhill under the force of gravity. The flows buried large areas covered by biosphere.

7 How have Earth's spheres changed since the eruption? Describe current interactions among Earth's spheres.

Vegetation is recovering, reappearing in areas where it had been wiped out. The geosphere is more stable because pressure was released by the eruption. The surface of the geosphere is still stabilizing as plants begin to hold material in place and erosion begins to smooth out steep surfaces. Geosphere materials suspended in the atmosphere are gradually falling to Earth's surface.

8 Predict how Earth sphere interactions in this region will change in the future.

Recovery of the biosphere and stabilization of the geosphere will probably continue until the geosphere erupts again. The next eruption will presumably have consequences similar to those brought about by the 1991 eruption—wiping out vegetation and animals, adding materials and energy to the atmosphere, and changing the shape of the geosphere. The cyclical process of the geosphere covering over the biosphere and the biosphere eventually covering back over the geosphere will probably continue.

Teacher's Guide Unit 3 Internet Investigation

Unit 3 Project Description

How can one volcano change the world? Select a volcano to study; use Web-based resources to gather information and develop a presentation illustrating the effects of eruptions from this volcano on the Earth system. Describe short-term and long-term changes to local and global Earth sphere interactions caused by this volcano. Present your findings in a format approved by your teacher. Your project should include all components listed below.

Unit Project Criteria

1. **A description of the plate tectonic setting of the volcano. (20%)**
   - Provide a map showing plate boundaries, including other volcanoes in the region.
   - Draw a cross-section diagram of the subsurface geology.
   - Describe the volcano's eruption history and magma characteristics.

2. **A description of the Earth system interactions affected by the eruption. (30%)**
   - Identify and describe how the eruption affected the Earth system on a local and global scale.
   - Explain the mechanisms by which Earth sphere interactions are altered by the volcano.
   - Provide supporting images, maps, charts, graphs, or animations.

3. **An explanation of the scientific studies, data, and images used to document changes caused by the volcano. (30%)**
   - Summarize any relevant scientific studies, data, images, or maps which provide evidence of Earth system changes caused by the volcano.

4. **A presentation of research and results. (20%)**
   - Communicate findings and ideas clearly and accurately.
   - Present appropriate data.
   - Use appealing visuals and design.

Earth System Interactions

Focus on the connections among Earth's spheres in your research and presentation. Think about the implications of volcanic eruptions on our planet as a whole. For example, consider:

- Explosive volcanic eruptions (geosphere) alter global climate (atmosphere) patterns, which affect plants and animals (biosphere) on land and in the oceans (hydrosphere).

- Plate tectonic processes shape the surface of the geosphere, controlling the shape and location of the hydrosphere and producing a variety of habitats for the biosphere.

- Human settlement patterns (biosphere) are influenced by volcanoes (geosphere).

**Teacher's Guide Unit 3
Internet Investigation**

4 UNIT

Earth's Changing Surface

NOAA

Measure the speed of a moving glacier

Compare before-and-after locations of stakes in the surface of a glacier. Calculate the speed of these flowing rivers of ice.

Investigate floods on the Mississippi River

Examine human attempts to control flooding along the Mississippi River. Check how well flood structures worked during major floods over the past century.

Examine storm damage along coastlines

Document the devastation caused by a hurricane on the east coast, and by strong El Niño-related storms on the west coast.

Unit 4
Earth's Changing Surface

UNIT 4 Earth's Changing Surface

OTHER WEB RESOURCES

VISUALIZATIONS

Spark new ideas for investigations with images and animations, such as:
• Meandering rivers
• Erupting geysers
• Retreating glaciers
• Seasonal migration of snow line

DATA CENTERS

Extend your investigations with current and archived data and images, such as:
• River maps
• Aquifers
• Glaciers
• Global wind patterns

EARTH SCIENCE NEWS

Relate your investigations to current events around world, such as:
• Flooding
• Coastal erosion
• Active glaciers
• Major dust storms

Unit 4
Earth's Changing Surface

When Is Mud Dangerous?　　▶ ES1204

1 **What conditions might cause mud to flow?**
Answers will vary. Angle of a slope or steepness of a hillside, degree of saturation by water, and volcanoes are all possible answers.

2 **What conclusion can you draw from the two images?**
The steeper the slope, the more likely that mud will slide.

3 **List at least two ways you could make the mud slide off the 30° slide plane without changing the plane's angle.**
Answers will vary. Adding more water and shaking the board are possible answers.

4 **What conditions in nature would be represented by the answers you gave for question 3?**
Answers will vary. Adding water would represent rain or flooding. Shaking the board would represent an earthquake.

5 **List at least two factors that contribute to the formation of mudflows on volcanoes.**
Answers will vary. Large amounts of loose ash, steep slopes of a volcano, and presence of water in lakes or glaciers are possible answers.

6 **How might forest fires affect an area's potential for experiencing mudflows?**
By destroying vegetation that protects soil from washing away, forest fires increase the potential for an area to experience a mudslide.

7 **Hypothesize about how mudflows could change the topography of an area after a fire.**
Answers will vary. Without vegetation to protect the surface, soil and mud will move downhill more easily, resulting in a general flattening of the topography.

8 **What human activities strip soil of its protective vegetation and increase its vulnerability to mudflows?**
Answers will vary. Deforestation and the clearing of natural vegetation for development are possible answers.

9 **Write a paragraph describing the conditions that cause dangerous mudflows. Include the types of locations where mudslides are most likely to occur.**
Answers will vary. Complete answers should discuss the following factors: slope angle, level of saturation by water, earth shaking, volcanic ash, glacial outbursts, and forest fires. Locations most likely to experience mudflows include canyon bottoms, stream channels, areas near the outlets of canyons or channels, volcanoes, and areas downstream from deforested land.

How Does Soil Vary from Place to Place?   ▶ ES1206

1 Write three observations about the soil in this photo.

The soil is very dark, probably fertile because of the existence of crops, well broken-up, and not in big clods.

2 Identify and describe as many different layers (horizons) as you see in this soil profile.

There seem to be four layers—a dark layer at the top with plants, a light gray layer which is thinner than the top dark layer, a very thick red-brown layer, and a bottom layer that is rocky and slightly darker.

3 List several similarities and differences across these soils.

The soils have different colors; some have defined layers; some have large pebbles included; some seem to be finer-grained.

4 What relationship do you think might exist between average annual precipitation and topsoil depth?

Answers will vary. Students may suggest that more precipitation relates to deeper topsoil, or vice versa, or that there is no relationship at all.

5 Measure and record the topsoil depth in your table.

	Topsoil depth in images (inches)	Average Annual Precipitation (inches)
Arizona	1	7
Montana	2	48
Pennsylvania	6	12
Massachusetts	8	43
Georgia	9	47
Hawaii	10	150

6 Sketch the graph on your worksheet.

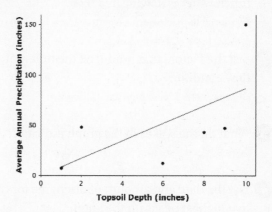

7 Describe any pattern that exists between topsoil depth and average annual precipitation.

Some of the data points appear to form a trend (shown on the graph) of increasing topsoil depth with increasing precipitation.

8 Go to the Web site for your state soil and estimate its depth.

Answers will vary with location. For example, the topsoil depth in Maryland is approximately 9 inches.

9 Go to the Web site for the United States precipitation map and find the average annual precipitation in your state.

Answers will vary with location. For example, the average annual precipitation in Maryland is 45 inches.

10 Plot the point for your state on the graph of topsoil depth versus precipitation. How does your state's soil compare to the others?

Answers will vary. For example, the plot for Maryland falls along the same trend as several other states.

11 Do average precipitation amounts appear to be related to the depth of topsoil in an area? If so, describe how. If you see no evidence for the relationship, suggest another factor you could plot versus topsoil depth to look for a relationship.

Answers will vary. Students may describe a relationship or suggest another factor for plotting.

How Does Stream Flow Change over Time?   ▶ ES1301

1 Describe three factors that control how much water is flowing in a river.

Factors include release of water from dams, amount of precipitation, and melting of snow.

2 List the factors that produced the different flow conditions.

Heavy rainfall; release from dam; snowmelt.

3 What do the spikes in the graph indicate?

Increases in the amount of water flowing in the stream.

4 List the date and amount of discharge for the two highest peaks on the graph.

August 10: 3 cfs: August 16: 1 cfs

5 What do the spikes in the graph indicate?

Heavy rains.

6 List the date and amount of precipitation for the two highest peaks on the graph.

August 10: 35 mm; August 15: 27 mm

7 Record the amount and dates of the five discharge spikes and amount and dates of the seven precipitation spikes. What do you notice about the timing of the peaks of discharge and rainfall?

Discharge peak dates (amount): 8/2 (2 cfs); 8/10 (3 cfs); 8/16 (1 cfs); 8/28 (<1 cfs); 8/30 (<1 cfs)

Precipitation peak dates (amounts): 8/2 (17 mm); 8/7 (13 mm); 8/10 (33 mm); 8/13 (2 mm); 8/15 (28 mm); 8/27 (14 mm); 8/29 (8 mm)

Peaks in precipitation precede increases in discharge.

8 What do you notice about the timing of the peaks of discharge and temperature variation starting in March? What do you conclude from this relationship? What other data might you need to confirm your hypothesis?

The peaks of discharge occur at the same time that temperatures start to go above 32 F. Discharge peaks may be related to the melting of snow. Data on the depth of snow would help confirm this relationship.

9 What do you notice about the timing of the peaks of discharge and rainfall? What do you conclude from this relationship?

Discharge rises and falls regularly. Rainfall peaks have no effect on discharge.

10 What relationship exists between temperature and discharge? What do you conclude from this relationship?

There is no relationship between temperature and discharge. Temperature does not control this discharge.

11 What do you think is responsible for the regular shape of the discharge peaks? Speculate on the cause of the major discharge peak at the end of March.

The discharge peaks are due to controlled releases of water from the Glen Canyon dam. Their regular spacing reflects more water releases during the weekdays when demand for electricity is high. On weekends, the discharge decreases. The major peak represents a simulated flood,

12 Write a summary describing the factors that affected the discharge of Hubbard Brook in February, 2000. Refer to specific peaks on the graphs.

The precipitation peak on 2/11 did not increase discharge. The temperature was below 0°C and snow depth increased, indicating that the precipitation was snow. The discharge peak on 2/15 was due to precipitation on 2/14. Precipitation on 2/16 and 2/18 produced no discharge peaks: these events were also snow. The discharge peak on 2/25 was due to a combination of precipitation on 2/24 and melting snow as indicated by temperatures above 0°C. The snow pack decreased from ~ 80 cm to ~60 cm. The large discharge peak on 2/28 was also due to precipitation and melting of snow, as indicated by the high temperature and decrease in snow pack depth.

What Controls the Shape of a Delta?   ▶ ES1304

1 Make a sketch of the main characteristics of each delta; then list them. Label the delta as wave-, river-, or tide-dominated.

<u>Mississippi River Delta</u>
River-dominated
- "Bird's Foot" shape
- Large sediment load controls shape
- Little wave or tide action

<u>Ganges-Brahmaputra Delta</u>
Tide-dominated
- Lobes perpendicular to shore
- Tides redistribute sediment

<u>Nile River Delta</u>
Wave-dominated
- Smooth coastline
- Waves redistribute sediment

2 Make a sketch of each delta. Identify the delta as wave-, river-, or tide-dominated.

Rhone: Wave-dominated

Tigris: Tide-dominated

Niger: Wave-dominated

Selenga: River-dominated

Gurupi: Tide-dominated

Fraser: River-dominated

Have Flood Controls on the Mississippi River Been Successful?   ▶ ES1308

1 Describe how each structure or activity illustrated in the images may reduce the impact of floods.

Levees build up banks so the river can hold more water. Canals help divert floodwaters away from agriculture or populated areas. Revetment involves laying wires along banks to stabilize the soil. Dredging deepens rivers by removing sediment, therefore allowing more water to flow in the channel. Snag boats remove large objects, like trees, that can impede flow in the river. Dams can hold and gradually release excess water, therefore decreasing the degree of flooding.

2 Briefly summarize how floods cause property damage.

Answers will vary. Floods can cause damage to property by depositing large amounts of mud and sand in buildings, destroying structures, and inundating crops.

3 Identify the events surrounding the Great Flood of 1927. Use the images and note some of the effects of the flood. Were attempts at flood control effective during this event?

The Great Flood of 1927 occurred as a result of excessive rains during the fall of 1926 and winter of 1927. Swollen rivers broke through several important levees, stranding hundreds of thousands of people. Some of the effects include flooded streets, broken levees, damaged cropland, and ruined homes. Efforts at flood control were not effective.

4 How was the flood control system constructed under the Flood Control Act of 1928 different from earlier attempts at flood control?

The Flood Control Act of 1928 provided for building of a lengthy levee system, more dams and locks, reservoirs to capture and hold floodwaters, and runoff channels to divert floodwater.

5 Examine the images and summarize the results of the 1937 flood of the Mississippi River. Were attempts at flood control effective in taming the 1937 flood?

Results of the 1937 flood included flooded streets, flooded farmland, destroyed roads, damaged and destroyed homes, and refugees. Flood control efforts do not appear to have been effective.

6 Describe some of the damage done by the flood of 1993. What contributed to such a disastrous flooding event?

Damage included flooded bridges and roads and homes, ruined cropland, houses swept away by flood waters, and damage to thousands of miles of levees. The major factor which contributed to this disaster was heavy rain during the summer of 1993 following spring melting of a very large snow pack. High soil moisture caused much of the rain to run off directly into rivers, resulting in extreme flooding.

7 Examine the three satellite images. Describe the major differences in the landscape before, during, and after flooding.

During the flood, the width of the river increased, covering large tracts of farmland. The Mississippi and Missouri Rivers overflowed their banks, especially near bends in the river. Receding floodwaters left behind barren land and some tracts of land that were still underwater.

How Does Water Move through the Ground?   ▶ ES1401

1 What characteristics of a rock layer allow it to hold water?

A rock layer would need to have spaces between grains or cracks or hollow spaces to hold water.

2 Record the volume of water pumped and the time it took to pump it. Calculate the flow rates.

	Volume	Time	Rate
Shale	1 L	4350 years	0.00023 L/yr
Sandstone	1 L	3.76 hours	0.27 L/hr
Limestone	1 L	43 years	0.023 L/yr

3 What differences do you see in the material of each rock layer? Describe particle size, particle shape, and the general shape and number of pore spaces in each rock.

Sandstone has sand-sized particles that are relatively round with many pore spaces and good connection between the pore spaces. Shale has smaller, flatter particles than sandstone. The pore spaces are more numerous, but smaller than in sandstone, with less area for water to move between the spaces. Limestone has no particle structure. It provides space for water storage only if cracks and cavities have been dissolved in the rock.

4 Describe how the shape and number of pore spaces affect permeability of these three rock types.

The more pore space there is, the more water can be held. If the shape of the pore space allows easy movement between pore spaces, the permeability will be higher.

5 Explain why the sandstone had the highest flow rate.

Water moves easily between the large, interconnected pore spaces.

6 Shale is composed of individual particles with pore spaces between them, similar to the structure of sandstone. Why was the flow rate for shale so much lower than for the sandstone?

Particles that make up shale are somewhat flat. They pack together more tightly than rounded sand grains. The resulting pore spaces are small and not well interconnected.

7 How could limestone's ability to hold and transfer water be increased?

If the voids and cracks in limestonewere to become larger and better connected as water dissolves away rock material, it could hold more water.

8 Discuss the suitability of sandstone, limestone, and shale as aquifers for supplying human water needs.

Shale makes a poor aquifer. Water flows through shale so slowly that it is considered impermeable. Sandstone makes a very good aquifer, because it generally has high porosity and permeability. Limestone can also be a good aquifer if it has high porosity and permeability, though the water would contain more dissolved minerals.

Teacher's Guide Chapter 14 Internet Investigation

How Many People Can an Aquifer Support?   ▶ ES1406

1 From 1884 to 1960, by how many feet did the water level in the Arapahoe aquifer drop?

Answers may vary.

5280 - 4880 = 400 feet

2 Give two explanations that could account for the decrease in the aquifer's water level.

Answers will vary. The aquifer could have received less recharge from precipitation or had more water discharged for human use.

3 Describe how changes in the climate of this region might affect the water level of the aquifer.

If the climate were wetter, the water level would rise. If the climate were drier, the water level would drop.

4 How might human activities affect the water level of the aquifer?

If people take more water out for farming and public water supply, the water level could drop. If people take less water from the aquifer, the level would rise.

5 How does the graph of precipitation compare to the decline in water level in the aquifer through the 1900's?

Precipitation rose slightly, while the water level of the aquifer dropped.

6 How did the population of the Denver area change between 1900 and 2000?

Population increased dramatically, from about 200,000 to 2,900,000 people.

7 Rank the four types of water use from the most water used to the least.

Irrigation, Industry, Public, Livestock

8 As population increases, new homes and businesses are being built around the margin of the Denver metropolitan area. Which of the current water use categories might increase in the future? Which might decrease?

Public and industrial uses might increase. Agriculture and livestock might decrease.

9 Hypothesize about what might have changed to bring the aquifer water level back up again after 1960. Describe how you might test your hypothesis.

Answers will vary. Students may mention more surface water being used, less water used for farming, conservation measures, or artificial recharge. To test their hypotheses, students would need to obtain water-use data or reports of water sources used over time. To support their hypothesis, data would need to show a decrease in discharge or an increase in recharge.

10 Calculate an estimate of the number of people who could get all the water they would need all their lives from the Denver Basin aquifer system. Assume that the aquifer contains 15 trillion gallons of water, people use 150 gallons of water per person per day, and they each live eighty years. Show your work.

Answers may vary.

15,000,000,000,000 gallons / (150 gallons/person/day x 365 days/year x 80 years) = 3,000,000 people

Teacher's Guide Chapter 14 Internet Investigation

Copyright © McDougal Littell Inc.

54 *Internet Investigations Guide*   ***Earth Science***

How Does Land Cover Affect Global Temperature?

1 Describe how the temperature on Earth would vary if it had a uniform surface.

A global temperature map would show bands of equal temperature parallel to the equator. Temperatures would decrease going from the equator to the poles.

2 Rank the four images in order of how much sunlight each reflects, starting with the largest amount of reflected light.

The glacier, the desert, the rain forest, the ocean.

3 For each of the three areas highlighted in the image, describe common land cover you would expect to find there. Note how it affects albedo.

The area that has a lot of vegetation has a low albedo; the area that has a lot of desert and exposed soil has a medium albedo; the area covered by snow/ice has a high albedo.

4 Compare global albedo values to global temperature patterns. Describe any pattern you observe.

Higher albedo generally corresponds to colder temperatures.

5 What happens to solar energy that is not reflected from the surface?

It is absorbed by the surface of Earth.

6 Explain how the energy that is not reflected affects the air temperature directly above the surface.

Energy not reflected is absorbed by Earth's surface and radiated back into the atmosphere, causing air temperature to rise.

7 Describe seasonal changes in the area covered by snow and ice.

During the winter, when ice and snow cover more sea ice, more light energy is reflected, reducing atmospheric temperature. During the summer, when open water covers more area, more light energy is absorbed, increasing atmospheric temperature.

8 Areas with high albedo absorb less energy than areas with low albedo do. Describe how increasing the portion of Earth's surface covered by snow and ice might change global temperature.

If the percent of land area covered by snow and ice were to increase, it would contribute to a reduction of average global temperature.

9 If a glacier melts back (covers less area), would the resulting changes in albedo tend to cause temperatures to rise or fall? Explain.

The changes in albedo would cause atmospheric temperature to rise, as more energy is absorbed by Earth's surface and radiated back into the atmosphere.

10 During the last ice age, glaciers covered a much larger area of Earth than they do today. Describe how the increase in albedo caused by more glaciers would have affected Earth's global temperature.

During the last ice age, the increase in area covered by glaciers caused lower atmospheric temperature. The lower atmospheric temperature allowed more glaciers to exist, further reducing temperature.

How Fast Do Glaciers Flow?

1 How could you measure the speed at which a glacier flows?

Answers will vary. Students may suggest comparing the location of an object on the glacier at different times.

2 Predict which stakes you would expect to move the farthest in one week. Explain your answer.

Answers will vary. The flags in the center of the glacier will move faster because there is less friction on the glacier in the center.

3 Which stakes moved the farthest?

Center flags.

4 These data show how far the flags moved in one week. At this rate, calculate how far the glacier would flow in 52 weeks (1 year).

In the center, the glacier would move about 14 meters 27.1 cm x 52 weeks = 1409 cm ≈ 14 meters

5 Why do you think the center of the glacier moves faster than the sides?

Less friction in the center of the glacier.

6 Describe what the movement of the Mathes Glacier might feel like if you were camped in the middle of it.

Answers will vary. The actual movement is so slight that you wouldn't be able to detect it without sensitive surveying instruments.

7 What are the highest flow rates observed on the Lambert Glacier? Which part of it is moving the fastest?

Highest flow rates are near 130 meters per year at the bottom of the glacier.

8 Earlier, you calculated a flow rate for the Mathes Glacier. The data collection site was near the beginning of the glacier, on one of its tributary branches. Check the flow rate of the Lambert Glacier in a similar location, near one of its tributary glaciers. How do flow rates of the two glaciers compare?

If you measure the flow rates in similar locations, they are quite similar—about 15 meters per year.

9 Explain why measuring flow rate at just one point along a glacier is not a good indicator of its overall flow rate.

The flow rate along the length of the Lambert Glacier varies considerably. If you only measured at one point, that location could be considerably faster or slower than the overall flow rate.

Earth Science

What Controls the Shape and Motion of Sand Dunes?

1 Which of the images shows sediments in motion? What is causing it to move?

Image A shows sediment in motion. Wind is causing it to move.

2 What accounts for the absence of dunes in the image to the right?

Dunes are absent because of the large-grain sediment, which cannot be transported by wind.

3 Why are dunes forming against the hay bales but not in the open field?

The hay bales stop sand from moving, causing a pileup of sand. In the open field, no obstacles are present.

4 Which of the images was taken in an area with abundant water?

Image A represents an area with abundant water.

5 Suggest two reasons for the presence of dunes in Image B but not in Image A.

Dunes are forming in Image B because the area is dry (no vegetation) and has an abundance of sand. Vegetation in Image A stabilizes the soil and prevents grains from blowing.

6 From these images, predict the conditions required for dune formation.

Dunes form in arid areas where there is an adequate supply of sand-sized sediment. Some obstacle is necessary for sand to pile up and begin dune formation. A transport mechanism like wind is also required for dunes to form.

7 In the boxes, sketch each dune type and indicate the wind dynamics and sand supply required to form each dune.

Barchan Dune	Longitudinal Dune
See page 343 in textbook for sketches Strong, steady winds Limited sand supply Curved ends point downwind	Winds shift slightly Moderate sand supply Ridges are parallel to wind
Parabolic Dune	**Transverse Dune**
Winds steady Abundant sand supply formed around blowouts Open ends point upwind	Winds steady Abundant sand supply Ridges are perpendicular to wind

8 Which planets or moons have dunes? State evidence to support your conclusions. What does the presence of dunes indicate about conditions on a planet or moon?

Mars has dunes. Images of the surface show abundant loose sediment and boulders and features that resemble barchan dunes on Earth. The presence of dunes on a planet indicates that it has loose, sand-sized sediments and a dynamic atmosphere to shape the sediments into dunes.

Where Did This Sand Come From?   ▶ ES1607

1 Describe the major characteristics of each sand grain. Your description should include details about shape, color, and surface features. Make a hypothesis about the source rock for each grain.

Answers will vary.

Sample 1:

Angular, dark, sharp edges, glassy

Volcanic glass

Sample 2:

Rounded, dark red, smooth

Magnetite

Sample 3:

Angular, light brown, irregular Bumpy surface

Feldspar

Sample 4:

Oval shape, light green, smooth surface

Olivine (green volcanic glass)

Sample 5:

Shaped like shell, white to tan, smooth

Shell fragments

Sample 6:

Irregular, tan, holes Rough sphere, clear to glassy, smooth.

Coral and Quartz

Sample 7:

Oval shape, red color, holes

Red volcanic fragment

Sample 8:

Oval shape, black, holes

Black volcanic fragment

2 Use your descriptions of sand grains to identify the three grains in the image.

Grain 1 - Red volcanic fragment

Grain 2 - Black volcanic fragment

Grain 3 - Green volcanic glass, may be olivine

3 What is the source of this sand sample?

The sand sample is from Hawaii. A volcanic rock is the source.

4 Describe the sand composition for each image. List a possible source rock for each sample.

Answers may vary.

Sample 1:

Grain 1 - Black volcanic fragment; Grain 2 - Shell fragment; Grain 3 - Red coral; Grain 4 - Quartz

Source: Volcanic rock near coral reef

Sample 2:

Olivine Sand

Source: Olivine-rich volcanic rock

Sample 3:

Grain 1 - Shell fragment; Grain 2 - Quartz; Grain 3 - Shell fragment; Grain 4 - Coral

Source: Coral reef near quartz source

Sample 4:

Red volcanic sand grains

Source: Reddish volcanic rock

Sample 5:

Green volcanic glass; Black volcanic fragment; Red volcanic fragment

Source: Mixed volcanic source

Sample 6:

Volcanic glass

Source: Glassy volcanic rock

Sample 7:

Coral and shell fragments

Source: Coral reef

Sample 8:

Magnetite sand

Source: Magnetic volcanic rock

How Do Storms Affect Coastlines?

1 Identify several factors that control the shape and character of coastlines.

Factors that control the shape of coastlines include plate tectonic setting, local geology, strength of waves, strength of wind, storms, and human activities.

2 Compare the images taken before and after Hurricane Dennis. Describe changes that occurred at each location.
• Location 1:

Significant erosion of point bar

• Location 2:

Beach loss on both sides of the island and on point bar in background

• Location 3:

Loss of beach. Several new channels cut across barrier island.

• Location 5:

Loss of beach. New and deeper channels cut across the island. Land in background is flooded.

• Location 7:

Loss of beach. New channels cut across barrier island.

3 What do you think happened to the sediments that are missing from these locations?

Answers will vary. Sediments have been redistributed along the shoreline or ocean bottom.

4 Compare the images of coastal areas before and after storms associated with El Niño, in the winter of 1997–98 . Describe changes that occurred at each location.
• Ventura, California:

Dramatic change in beach area. More sand in mouth of bay.

• Point Reyes, California:

Increase in amount of sediment deposited. Beach area is larger. New sandbars in channel are also present.

• Cape Blanco, Oregon:

Beach is larger. New channel cuts beach.

5 In these photos, where do you think the sediments that increased the beach areas came from?

Answers will vary. Sediments have been redistributed from the ocean bottom onto the shore.

6 Describe how the atmosphere, hydrosphere, and geosphere interact to affect coastlines.

Answers will vary. Storms in the atmosphere affect the hydrosphere, which churns up geosphere materials along coasts and redistributes them along the shore or ocean bottom.

What are the Costs and Benefits of Damming a River?

1 List three positive and three negative aspects of damming rivers.

Positive aspects may include water availability, recreation, and hydroelectricity. Negative aspects may include pollution, loss of natural environments, and loss of cultural artifacts.

2 Describe environmental costs associated with human recreation at Lake Powell.

Costs include pollution and littering from humans and vehicles, crowding, and loss of wilderness areas.

3 What benefits do archeological sites and artifacts offer a culture?

They provide cultural connections, data on past civilizations, and records of climate change.

4 Describe how you might have felt about a lake flooding the canyon if you had lived in the Glen Canyon area.

Answers will vary. They may include empathy for Native Americans who lost their homes, or they may welcome the opportunity of new jobs in the area.

5 Describe how discharge levels and water temperature in the Colorado River below Glen Canyon Dam changed after the dam was built.

Answers will vary. Discharge rates now change with demand for electricity. The temperature of river water below the dam is now lower and more constant than before the dam was built.

6 Describe the importance of annual floods to canyon flora and fauna.

Annual flooding clears debris and deposits sediment and nutrients for plant growth.

After you explore the Web pages on each aspect listed in the table, evaluate the impact of that aspect as a cost or benefit. Use a scale from -2 to +2 to indicate your assessment.

Recreation	Answers will vary.
Native Culture	
Changes in the Water	
Hydroelectricity	
Irrigation	
Preservation	
Total Score	Answers vary from -12 to +12

7 Describe the former environment of Glen Canyon. What were some of the benefits this natural area offered?

Glen Canyon was a deep sandstone canyon with streams and trees. It provided a relaxing environment for humans to connect with nature, and habitat for native species of plants and animals.

8 Does your total score accurately represent your overall opinion of the impact of Glen Canyon Dam and Lake Powell? Describe any discrepancies.

Answers will vary. Students could calculate an overall positive score yet feel that the loss of the canyon environment was more important than all other aspects.

9 How could you change the rating system to indicate that some aspects are more important than others?

Answers may vary. Students could apply weighting factors to the aspects before summing up the total score.

10 What other aspects of the dam and lake might you rate?

Answers will vary. Students may mention opportunity for jobs, loss of water due to evaporation and infiltration, or downstream effects on river ecosystems.

What are the Costs and Benefits of Damming a River?

Unit 4 Project Description

How are Earth sphere interactions altered by the damming of rivers? Using Web resources, identify a major river that has been dammed. Obtain information and scientific studies on the dam and how it has affected the river system and changed interactions among Earth's spheres. Present your findings in a format approved by your teacher. Your project should include all components listed below.

Unit Project Criteria

1. **A description of the river system and the dam. (20%)**
   - Describe the entire river system. Include a general description of the landscape through which the river flows, the major sources of its water, annual flow patterns, plant and animal ecology, and human uses.
   - Provide a description of the dam and information about why the dam was built.
   - Develop a descriptive map with representative images of the river system.

2. **A description of changes to the river system resulting from the dam. (30%)**
   - Identify and describe changes to the river system caused by the dam.
   - Provide supporting images, maps, charts, or graphs to illustrate these changes.

3. **A summary of scientific studies on the changes caused by the dam. (20%)**
   - Summarize studies and information about the changes caused by the dam.
   - Summarize pertinent studies from different rivers and explain how the findings from these studies might be applied to this river.

4. **An evaluation of the changes caused by the dam. (10%)**
   - Evaluate each of the identified changes as positive or negative.
   - Present appropriate information that supports these ratings.

5. **A presentation of research findings. (20%)**
   - Communicate findings and ideas clearly and accurately.
   - Use appealing visuals and design.

Earth System Interactions

Focus on the connections among Earth's spheres in your research and presentation. Think about how the damming of rivers affects our planet as a whole. For example, consider:

- Dams change river flow dynamics, subsurface water flow patterns, and water quality. Biosphere access to the hydrosphere is changed.

- Distribution of plants and animals (biosphere) changes radically due to flooding (hydrosphere) of habitats upstream of the dam and changes in in flow patterns downstream from the dam.

- Dams increase the surface of the hydrosphere exposed to the atmosphere. This results in increased evaporation into the atmosphere.

- Sediments (geosphere) normally transported downstream by free-flowing rivers are deposited in lakes, changing the dynamics of erosion and deposition along the river (hydrosphere).

5 UNIT Atmosphere and Weather

NASA

Track a hurricane
Get the latest satellite images of these massive storms and track their movement.

Make your own weather forecast
Use current maps of cloud cover and precipitation for your local area to forecast tomorrow's weather.

Investigate global climate change
Read the record of climate change recorded in Antarctica's ice. Compare today's changing climate with climate changes over the last 400,000 years.

Earth Science

5 UNIT Atmosphere and Weather

OTHER WEB RESOURCES

VISUALIZATIONS

Spark new ideas for investigations with images and animations, such as:

• Clouds forming and dissipating

• Weather systems

• Forest fires

• Auroras

DATA CENTERS

Extend your investigations with current and archived data and images, such as:

• Satellite images of cloud cover

• Radar images of precipitation

• Global and regional temperature maps

• Safety tips for hurricanes, tornadoes and snow storms

EARTH SCIENCE NEWS

Relate your investigations to current events around world, such as:

• Hurricanes

• Tornadoes

• Extremely hot or cold weather

• Forest fires

Unit 5
Atmosphere and Weather

What Can You Learn from a Thermometer on a Rising Balloon? ▶ ES1702

1 List at least three atmospheric properties that might change as you traveled from the Earth's surface to space.

Answers will vary. Students may suggest changes in clouds, temperature, pressure, ozone, and dust.

2 Write a general statement that describes how the air temperature changes as the balloon rises through the atmosphere.

The temperature increases for the first 1,000 to 1,500 meters (m) to a high of 24°C, then drops steadily for the next 11,000 m to a low of about –63°C. It fluctuates around –60°C for the next 4,000 m, then gradually rises to –43oC at an altitude of about 31,000 m.

3 Commercial airplanes typically fly at an altitude of about 12,000 meters. What would the air temperature be at that altitude?

Around –63°C

4 What is the range of altitudes for the top of the troposphere?

Approximately 10,000 m to 17,000 m

5 What is the range of the lowest temperatures found at the top of the troposphere?

–54°C to –66°C

6 How does the height of the tropopause change from the polar region to the equator?

The tropopause is lower near the poles and higher near the equator.

7 How does the minimum temperature of the tropopause change from the equator to the polar region?

Minimum tropopause temperatures are lower near the equator and higher near the poles.

8 In general, how does the height of the tropopause change from day to night?

The height of the tropopause is higher at midday (16,000 m) than at midnight (14,000 m).

9 Make a general statement about what happens to the speed of the wind as the balloon rises through the atmosphere.

Wind speed increases steadily from about 4 knots at the surface to about 85 knots at an altitude of about 13,000 m. Wind speed decreases sharply above 13,000 m, dropping to almost zero at an altitude of about 21,000 m. Above 21,000 m, wind speed starts to rise again.

10 How does the altitude at which the wind speed stops increasing and begins decreasing compare to the altitude of the change from troposphere to stratosphere?

They are approximately the same altitude.

Earth Science

1 With a small group of other students, brainstorm a list of factors that might influence the temperature of a location.

Answers will vary. Sample answers include latitude, elevation, proximity to water, cloud cover, humidity.

2 What do you think causes the systematic shift of radiation and temperature bands through the year?

Answers will vary. Sample answers include seasonal changes, angle of Earth to sun, and length of day.

3 Write a statement that describes the relationship between temperature and elevation.

Temperatures are higher at lower elevations and lower at higher elevations.

4 Look at the trend line through the scatter plot. What is the approximate change in temperature for every 1,000 foot (ft) increase in elevation?

Temperatures drop about 3°F for every 1,000 ft increase in elevation.

5 Shrine Pass is about 75 miles west of Denver at an elevation of 11,050 ft above sea level (about 6,000 ft higher than Denver). If the temperature in Denver is 82°, predict the temperature at Shrine Pass.

About 64°F.

6 What might be responsible for the observed difference in temperature between the coastal cities and the valley cities?

Answers will vary. Students may suggest the cooling effect of water, or that mountains prevent cooling breezes from moving into the valley.

7 Find the average temperature for marked cities in the central valley.

The average temperature in the valley was 90°F.

8 Find the average temperature for marked cities along the California coast.

The average temperature along the coast was 66.3°F.

9 What is the difference in average temperatures for cities in the central valley and along the coast?

The difference between the central valley and coastal temperatures is about 23.7°F.

10 Predict how the temperature difference between valley and coastal cities might change during winter months. Explain your reasoning.

Answers will vary. Students may suggest that the temperature difference will increase, decrease, or stay the same. Look for reasonable justification of their answer.

11 List at least one other factor that might control temperature. Describe how you might quantify it.

Answers will vary. Students may suggest latitude, wind, cloud cover, or humidity. To quantify the effect of the factor, they would compare average temperatures under varying conditions of the factor they suggest.

Teacher's Guide Chapter 17
Internet Investigation

1 During which months does the South Pole have the lowest concentrations of ozone? Estimate the minimum ozone level.

September through November. Minimum ozone concentrations are around 150 Dobson units.

2 During which months does the South Pole have the highest concentrations of ozone? Estimate the maximum ozone level.

January through March. Minimum ozone concentrations are around 300 Dobson units.

3 During which months does the North Pole have the lowest concentrations of ozone? Estimate the minimum ozone level.

August through September. Minimum ozone concentrations are around 300 Dobson units.

4 During which months does the North Pole have the highest concentrations of ozone? Estimate the maximum ozone level.

February through April. Maximum ozone concentrations are around 500 Dobson units.

5 Describe how ozone concentration changes over the North and South Poles through a year. Compare minimum ozone levels for each pole.

Answers will vary: The North Pole has high levels of ozone at the end of its winter, and low levels at the end of (northern hemisphere) summer. The South Pole has high levels of ozone during the southern hemisphere's late summer-early fall. Low levels of ozone occur during the southern hemisphere's spring and summer. Minimum ozone concentrations over the South Pole range from 150-300 Dobson units. Minimum ozone concentrations over the North Pole range from 300-500 Dobson units.

6 Record the length of the major and minor axes of the ozone hole and the minimum ozone level for each year displayed.

Year	Major Axis (km)	Minor Axis (km)	Hole Area (km²)	Ozone Level
1979	0	0	0	280
1981	2500	1700	3,300,000	220
1983	3700	3200	9,300,000	180
1985	4400	4200	14,500,000	160
1987	5600	4600	20,000,000	140
1989	5500	4400	19,000,000	150
1991	5600	4500	20,000,000	130

7 Calculate and record the area of the ozone hole for each year displayed.

8 On a separate sheet of paper, sketch and label your "hole area versus time" graph.

9 Describe the overall trend of the size of the ozone hole from 1979 to 1991

No hole was observed in 1979. The size of the hole increased steadily through 1991.

10 On a separate sheet of paper, sketch and label your "minimum ozone level versus time" graph.

11 Describe the overall trend of the minimum ozone level from 1979 to 1991.

The minimum level dropped from about 280 Dobson units in 1979 to about 130 Dobson units in 1991.

12 How does the graph of minimum ozone levels in the late 1990s compare to the graph of levels before that time?

Since the early 1990s, the minimum ozone levels have been rising.

Which Way Does the Wind Blow?   ▶ ES1806

1 Make a prediction: Which way do you think the wind blows over Oahu?

Answers will vary. Students should answer by using the direction from which the wind blows—for example, from the northeast.

2 What evidence did you use to make your prediction?

Answers will vary. Students may cite the location of clouds or the smooth water southwest of the island.

3 Describe the pattern of vegetation you see on Oahu. In which areas are plants concentrated? Which areas are bare or have little vegetation?

The mountains are covered in vegetation. The low areas southwest of the mountains look dry, with little vegetation.

4 Describe what happens to moist air as it moves up the slope of a mountain, crosses the summit, and moves back down the leeward slope.

As air rises, it cools. Clouds form and rain starts. Once the air moves over the summit and begins moving down the opposite slope, the air warms, the rain stops, and clouds dissipate.

5 How would vegetation patterns reflect the rainfall amounts on each side of a mountain where wind usually blows in one direction?

The rainfall on the windward side of the mountain where air is rising should have a lot of vegetation, while the dry side is where the air is descending from the summit, and there will be less vegetation.

6 What does the vegetation pattern on Oahu suggest about the wind direction?

The northeast side of the island has heavy vegetation. This suggests that the wind blows from the northeast to the southwest.

7 What are the three places with the highest annual precipitation levels, and where are they located on the island?

Manoa (37.8 in.), Nuuanu (35.9 in.); and Wilson and Ahuimanu (31.9 in. each) are located near the south end of the mountain range that forms the northeastern part of the island.

8 What are the three places with the lowest annual precipitation levels, and where are they located on the island?

Honolulu (8.0 in.), Waianae (8.6 in.), and Aloha Tower (9.3 in.) are along the southwestern coast of the island.

9 Describe at least one way you could check if your predicted wind direction is correct.

Answers will vary. Students may suggest checking with the National Weather Service on the Internet or watching weather reports on T.V.

10 If a weather report for Oahu indicated a storm with winds from the south, would this refute the evidence you gathered about wind direction? Why or why not?

No. The vegetation and rainfall patterns indicate the long-term general wind direction, not the daily weather.

11 Drawing on what you learned from the case of Oahu, in what direction do you predict the wind blows over California? Describe the evidence you used to make your prediction.

The wind blows from the west. The heaviest rainfall is on the west side of the the Sierra Navada Mountains, and the east side is very dry.

Copyright © McDougal Littell Inc.

Teacher's Guide Chapter 18
Internet Investigation

1 List at least three things that you currently know about acid rain.

Answers will vary. Students may know that burning fossil fuels causes acid rain.

2 Which parts of the country experience the lowest pH values? Which part experiences the highest pH values?

Lowest pH: Great Lakes area, New England, Ohio, New Jersey, and Pennsylvania. Lowest pH: the western half of the country.

3 Over the six year period, did the acid rain problem increase or decrease? Explain your answer.

Decreased—areas of lowest pH had increases in pH levels, indicating less acid.

4 Describe the acid rain situation for your own area. How has it changed over the six years?

Answers will vary. Levels of pH increased, decreased, or stayed the same.

5 Which areas have the highest concentrations of sulfate ions (SO_4^{2-}) in rainwater?

The areas around Ohio and Pennsylvania, southern Illinois and Indiana, and western New York

6 How do the locations of fossil fuel-burning power plants compare to the location of areas with high concentrations of sulfate ions?

The areas with high concentrations of sulfate ions are generally east (downwind) of coal plants.

7 How do the areas of high nitrate ion (NO_3^+) concentrations correspond to areas of high sulfate ion concentrations?

The areas with high sulfate ion concentrations generally also have high nitrate levels. Areas in the central and southwestern U.S. also show elevated nitrate levels.

8 What might account for high nitrate ion concentrations in areas that have low sulfate ion concentrations?

Answers will vary. The presence of power plants and population centers are strong candidates.

9 What might account for high ammonium ion (NH_4^+) concentrations in the central portion of the United States?

Agriculture

10 Explain how sulfate, nitrate, and ammonium levels work together to produce the observed pH level in central Pennsylvania.

The area has high concentrations of sulfate and nitrate ions but low concentrations of ammonium ions. The low concentration of ammonium cannot counteract the other ions, so the overall pH level is low.

11 Explain how sulfate, nitrate, and ammonium levels work together to produce the pH level in the Panhandle region of western Oklahoma.

The area has moderate levels of sulfate ions and high levels of nitrate ions. The very high levels of the basic ammonium ions counteract the acids, so overall pH is high.

12 Record and graph ten years of NH_4^+, NO_3^+, SO_4^{2-} and Lab pH values recorded at a site near your home.

Answers will vary.

Year	NH4$^+$	NO3$^+$	SO4^{2-}	pH
1991	0.13	1.05	2.20	4.37
1992	0.18	1.40	2.16	4.36
1993	0.14	1.16	1.96	4.46
1994	0.18	1.19	1.78	4.39
1995	0.16	1.02	1.33	4.52
1996	0.13	0.93	1.37	4.52
1997	0.20	1.43	1.90	4.35
1998	0.13	0.86	1.57	4.48
1999	0.13	0.98	1.67	4.47
2000	0.18	1.27	1.94	4.39

How Does the Jet Stream Change through the Year?

1 During which months is the jet stream located closest to the North Pole?

June, July, August, and September

2 During which months is the jet stream located closest to the equator?

December, January, February, and March

3 During which months does the jet stream have the highest wind speeds? How fast is the jet stream during these months?

December, January, and February. Jet stream wind speeds are 140 miles per hour.

4 During which months does the jet stream have the lowest wind speeds? How fast is the jet stream during these months?

July, August, and September. Jet stream wind speeds are 60-70 miles per hour.

5 On February 1, at what latitude do you find the fastest part of the jet stream over North America?

Approximately 35°N latitude.

6 On July 15, at what latitude do you find the fastest part of the jet stream over North America?

Approximately 45-48°N.

7 In your own words, describe how the jet stream changes in location and intensity over the course of a year.

Answers will vary. The jet stream is the fastest and the farthest south during the winter months of December, January, February, and March. It is the least intense and the farthest north during the summer months of July, August, and September.

8 What differences might you expect between the southern and northern hemisphere jet streams?

Answers will vary. Because the southern hemisphere seasons are the opposite of those in the northern hemisphere, the jet stream will be most intense and closer to the Equator during the months of June, July, and August. It will be least intense and at a higher latitude during the months of December, January, and February.

9 Sketch and label the average position of the summer jet stream. Do the same for the average position of the winter jet stream.

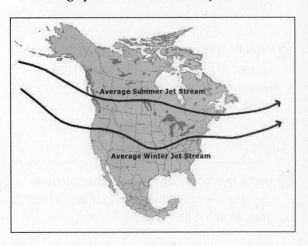

Could You Break the Record for an Around-the-World Balloon Flight?

▶ **ES1908**

1 **How do you steer a balloon?**

By moving either up or down in elevation, the balloon catches winds going in different directions.

2 **Which jet stream(s) offer the best path that meets the rules for an around-the-world flight?**

Either the northern or the southern subtropical jet stream.

3 **During which season do jet streams appear to offer optimal conditions for around-the-world balloon flight attempts?**

Winter (January for northern hemisphere, July for southern hemisphere.)

4 **Suggest two reasons why you might not want to use the northern polar jet stream for an around-the-world attempt.**

It takes a meandering path rather than a straight path. It is also associated with many storms, which might make it more dangerous for the balloonists.

5 **Which jet stream did the Breitling Orbiter 3 appear to follow? Which jet stream did the Solo Spirit 3 appear to follow?**

Breitling Orbiter 3: Northern subtropical jet stream.
Solo Spirit 3: Southern subtropical jet stream.

6 **Using the following information, calculate the average speed that the Breitling Orbiter 3 and the Solo Spirit 3 flew.**

Breitling Orbiter 3:
25,361 miles / 478 hours = 53 miles/hour.
Solo Spirit 3:
14,236 miles / 206 hours = 69 miles/hour.

7 **Which jet stream do you think offers the best chance for flying around the world? Explain your choice.**

Answers will vary. Look for justification for whichever jet stream students choose.

8 **Given this choice of jet streams, where would the best takeoff point be? Explain your choice.**

Answers will vary. Look for justification for whichever location students choose.

9 **What is the maximum speed at which this jet stream would carry the balloon?**

Answers will vary. Students will indicate maximum speed of the jet stream they choose.

10 **If this air speed could be maintained, how many hours would it take to fly 20,000 miles?**

Answers will vary.
20,000 miles / max. speed from #9 = number of hours.

11 **Based upon this information, do you think this balloon flight would beat Breitling Orbiter 3's record? Why or why not?**

Answers will vary.

How Does a Mid-Latitude Low Develop into a Storm System?

1 On which side of the cold front are clouds and precipitation generally found?

Clouds develop on the leading edge of the cold front, and precipitation occurs just behind it.

2 On which side of the warm front are clouds and precipitation generally found?

A wide band of clouds is found in front of the leading edge of the warm front. Precipitation occurs below this broad band of clouds.

3 What additional information would be helpful in determining the center of the low-pressure system?

Surface weather observations, especially barometric pressure, from many locations would show the position of the low pressure system.

4 Describe the weather conditions in each of the four cities on March 13, 1993 at 0200Z.

Students describe conditions as indicated on surface weather data tables.

Example: Asheville: Temperature 32° F, snow and fog, visibility 1 mile, wind 9 mph, wind direction 070°, 2 inches of snow on the ground.

5 What do the barometric pressures for Asheville and Concord tell you about the location of the storm relative to these cities?

Barometric pressure at Asheville is low, indicating the storm center is near; the Concord barometric pressure is still high, indicating that the storm is not as close.

6 How do the surface weather observations compare to the colors indicated on this satellite image?

Blue and purple show more intense weather than pink and yellow do.

7 Describe the weather conditions in each of the four cities on March 13, 1993 at 1000Z.

Students describe conditions as indicated on surface weather data tables.

Example: Asheville: Temperature 30° F, intermittent snow and fog, visibility 1 mile, wind 14 mph, wind direction 340°.

8 Describe how surface weather conditions changed as the storm approached, passed through, and moved away from these locations.

Temperatures and barometric pressures drop as the storm approaches; wind speeds are high as the storm arrives and snow falls; barometric pressures rise and winds die down as the storm passes.

9 Based on the pattern of storm movement and the surface weather data, predict the satellite image colors over each of these four cities on March 14, 1993 at 000Z.

Asheville-white or yellow; Wilmington-blue, pink, or purple; La Guardia-blue, pink, or purple; Concord-blue, pink, or purple.

How Well Can You Predict Tomorrow's Weather? ▶ ES2013

1 Where are low-pressure areas located on this map? Where is the high-pressure area located? Explain how you determined this.

There are two low-pressure areas; one is in the Pacific Northwest, and the other is over the Northeast. A high-pressure area can be found over the middle of the U.S. Barometric pressure is indicated on the yellow lines, with the lowest pressure in the middle of the lows.

2 What does this image tell about the weather near the low-pressure areas that the previous image does not?

It shows the location of high clouds and stormy weather near the low-pressure areas.

3 Describe the motion of clouds around the low-pressure area located over northeastern North America.

The clouds appear to be moving in a counterclockwise direction around the low-pressure area.

4 What kind of weather is predicted for the central part of the U.S.? For Oregon and Washington?

The high pressure in the middle of the U.S. will result in cloud-free weather. There will be rain in western Oregon, and a cold front moving through the middle of both states will cause a cooling trend.

5 Use your knowledge and experience plus Web resources to write a prediction for tomorrow's weather. Record your predictions in the table and describe the sources of your information.

6 Describe the meteorological processes you believe will most influence tomorrow's weather (e.g. cold or warm fronts, low or high pressure, air masses, etc.) in your region.

Answers will vary.

7 Day Two: Check your predictions. How accurate was your prediction? What weather factors did not behave as you had expected?

Answers will vary.

Weather Variable	Prediction	Source for Information
High temperature	Answers will vary.	
Low temperature		
Cloudiness		
Wind speed		
Wind direction		
Prediction (Type and amount)		
Barometric pressure		

Earth Science

What Factors Control Your Local Climate?

1 Why might one place have cold, snowy winters, while it rarely snows at another place only a hundred miles away?

Major differences in elevation could account for this.

2 Provide a justification for each factor you indicated as having major influence on the climates of Phoenix, AZ and Flagstaff, AZ.

Elevation is the major climate control factor. Higher elevations in Flagstaff explain lower temperatures as well as greater amounts of snow.

3 Summarize the major differences in climate between Boston, MA and North Bend, OR.

Temperatures in North Bend stay fairly constant year round, whereas Boston has greater temperature variations. Boston has snow in winter, while North Bend does not.

4 Provide a justification for each factor you indicated as having major influence on the climates of Boston, MA and North Bend, OR.

Ocean temperature and ocean currents differ greatly between Boston and North Bend.

5 Summarize the major differences in climate between Fargo, ND and Dallas, TX.

Fargo has much colder minimum temperatures and more snow than Dallas.

6 Provide a justification for each factor you indicated as having major influence on the climates of Fargo, ND and Dallas, TX.

Latitude appears to be the biggest influence.

7 On another sheet of paper, write a description of your local climate. Discuss characteristics such as timing and types of precipitation you receive, minimum and maximum temperatures and their timing, and any other aspects you think most accurately describe your local climate.

Answers will vary. Check for completeness.

8 Describe how each of the climate control factors in Data Table 2 influences your local climate.

Answers will vary. Check for descriptions of your local climate control factors.

Data Table 1: Comparative Climates

		Phoenix	Flagstaff	Boston	No. Bend	Fargo	Dallas	My Home
Locational Information	Elevation (ft)	1112	7004	30	10	895	574	
	Latitude	33.3N	35.1N	42.2N	43.2N	46.5N	32.5N	
	Longitude	112.0W	114.4W	71.0W	124.5W	96.5W	97.0W	
Temperature (Maximum)	Jan/July	65/108	40/82	38/80	50/65	15/82	52/95	
	Maximum	122	97	102	95	105	113	
Temperature (Minimum)	Jan/July	40/78	13/47	22/62	39/52	0/60	32/72	
	Minimum	19	-23	-7	13	-35	-1	
Precipitation (Rain)	Jan/July	.03/0.01	0.06/0.02	0.09/0.1	0.27/0.04	0.02/0.12	0.06/0	
	Average	7.7	22.9	41.5	63.3	19.5	33.7	
Precipitation (Snow)	Jan/July	0/0	0.06/0	0.02/0	0/0	0.03/0	Trace/0	
	Average	0	108.9	41.7	1.8	40.1	3.1	

Data Table 2: Rating the Influence of Climate Control Factors on Local Climate

Climate Control Factors	Phoenix	Flagstaff	Boston	No. Bend	Fargo	Dallas	My Home
Latitude	m	m	m	m	Major	Major	
Elevation	Major	Major	m	m	m	m	
Water	m	m	Major	Major	m	m	
Ocean Current	m	m	Major	Major	m	m	
Topography	m	m	m	m	m	m	
Prevailing Winds	m	m	Major	Major	m	m	
Vegetation	m	m	m	m	m	m	

Key: Major= Major influence **m** = Minor influence

Teacher's Guide Chapter 21 Internet Investigation

How Do Ice Cores of Glaciers Tell Us about Past Climates?

1 Over the past 165,000 years, how much has the temperature varied above and below the current average temperature?

From about 7° below to about 1° above.

2 Does there appear to be any correlation between the major high and low points on the two graphs? Explain, using specific examples.

Yes. High temperatures generally occurred near the same time as high CO_2 concentrations, and vice versa.

3 What do the data suggest about the relationship of temperature and the amount of atmospheric carbon dioxide (CO_2)?

Possible conclusions: 1) Increases in temperature cause increases in CO_2 concentrations; 2) Increases in CO_2 concentrations cause increases in temperature; and 3) Temperature and CO_2 concentrations parallel each other, but do not affect each other.

4 Which of the two variables, temperature or CO_2, appears to have begun rising first?

CO_2 began rising before temperature.

5 Which of these two variables appears to have peaked first?

CO_2 peaked first, at about 135,000 years ago. Temperature peaked about 133,000 years ago.

6 Which of these two variables appears to have begun dropping first?

CO_2 started falling before temperatures began to fall.

7 Examine the relationship between temperature and CO_2 for the period from 20,000 years ago to the present. Explain what you observe.

Answers will vary. CO_2 started rising just before temperature began to rise; temperature continued to rise after the rise in CO_2 began to slow.

8 Based on your observations, hypothesize about the relationship between temperature and the amount of CO_2 in the atmosphere.

Changes in CO_2 and temperature are directly related. Increases in CO_2 cause increases in temperature, and decreases in CO_2 cause decreases in temperature.

9 Based on your observations of CO_2 amounts, predict how temperatures changed from 1958 to 2001.

Temperatures should have increased as CO_2 amounts increased. The rise in temperature should begin after the rise in CO_2.

10 For the period from 10,000 to 30,000 years ago, describe the apparent relationship between temperature and the amount of dust in the atmosphere.

The temperature is low and the amount of dust is high.

11 Do data for the time period of 50,000 to 70,000 years ago confirm or refute the relationship you stated in question 10? Explain how.

Confirm, because dust amounts are high when temperatures are low.

12 Explain your reasons for drawing the temperature line as you did.

Answers will vary. Lines were drawn to show that dust amounts are higher when temperatures are lower.

How Might Global Climate Change Affect Life on Earth?

1 Why do you think the error bars become smaller in the more recent part of the graph?

Accuracy in recording and reconstructing temperatures has increased through time.

2 When did concentrations of these greenhouse gases start rapidly increasing? What might have caused these increases?

Around 1800. The industrial revolution brought heavy reliance on fossil fuels.

3 Which gas affects global warming most? Which gas has the least effect?

Most effect—CO_2. Least effect—NO_2.

4 Predict how continued increases in atmospheric carbon dioxide will affect global temperatures.

Increasing CO_2 concentrations will cause increasing global temperature.

5 Why do scientists develop numerous models rather than rely on just one?

Models represent best guesses based on observations. Using just one model might be inaccurate. Also, many models can be compared to each other.

6 Based on all the models, what are the minimum and maximum temperature increases expected to occur by 2100?

Average surface temperature is projected to increase by 1.4 to 5.8° C by 2100.

7 Which hemisphere (north or south) appears to have experienced the most significant temperature changes? Hypothesize about why this is so.

Northern hemisphere. Answers will vary. Water moderates temperature changes and there is more ocean area in the southern hemisphere. More fossil fuels are burned in the northern hemisphere.

8 Which global climate change impact do you think poses the greatest risk to humanity? Cite evidence for your answer.

Answers will vary. Look for evidence and reasonable justification.

9 Suggest two actions that people might take to decrease human influence on global climate change.

Answers will vary.

Climate Change Impact Assessment Table			
Impact Category	**Description of Expected Impacts**	**Most Vulnerable Locations**	**Ability of Humans to Adapt to Impact**
Health	Answers will vary.		
Agriculture			
Water Resources			
Forests			
Species and Natural Areas			
Coastal Areas			

How Might Global Climate Change Affect Life on Earth?

Unit 5 Project Description

How will processes in the Earth system change in response to global climate change? Using Web resources, identify an impact that is expected to occur as a result of global climate change. Develop a presentation that explains the impact and how it will change the Earth system. Present your findings in a format that has been approved by your teacher. Your project should include all components listed below.

Unit Project Criteria

1. **A description of the impact caused by global climate change. (20%)**
   - Explain the nature of the impact and its expected location, timing, and duration.
   - Use images, graphics, or animations to illustrate the impact.

2. **A description showing how global climate change affects the Earth system. (30%)**
   - Identify portions of the Earth system affected by climate change.
   - Describe how climate change alters the normal functioning of the system and leads to the impact.
   - Use images, graphics, or animations to illustrate the changes in Earth system function.

3. **An explanation of the scientific evidence linking the impact to global climate change. (20%)**
   - Describe the scientific evidence for the relationship between the impact and global climate change.
   - Identify strengths and any uncertainties in the scientific evidence.

4. **A recommendation for decreasing the degree of the impact. (10%)**
   - Explain ways that the impact could be averted or decreased.

5. **A presentation of research and results. (20%)**
   - Communicate findings and ideas clearly and accurately.
   - Present appropriate data.
   - Use appealing visuals and design.

Earth System Interactions

Focus on the connections among Earth's spheres in your research and presentation. Think about how the impact you are researching affects our planet as a whole. For example, consider:

- Global climate change observed in the atmosphere also affects processes in the biosphere (changes in distribution of species), hydrosphere (changes in circulation patterns), and geosphere (changes in weather patterns change erosion patterns).

- Release of greenhouse gases caused by humans burning fossil fuels (biosphere and geosphere) appears to be a major factor contributing to global warming (atmosphere).

- Oceans (hydrosphere) absorb and redistribute heat energy to the other spheres, leading to environmental changes which affect distribution of the biosphere.

6 UNIT Earth's Oceans

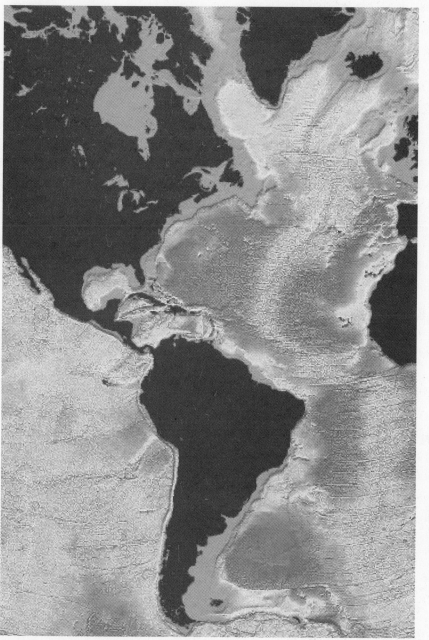

NOAA

Drain the ocean to explore its floor

Use visualization tools to drain the oceans, revealing mountain chains and deep trenches.

Chart a path to sail around the world

Use your knowledge of ocean currents and wind patterns to plot the fastest path for sailing around the world.

Find the best fishing spots in the Atlantic

Analyze sea surface temperature images of the Gulf Stream in the Atlantic Ocean to choose the best locations for fishing.

Unit 6
Earth's Oceans

OTHER WEB RESOURCES

VISUALIZATIONS

Spark new ideas for investigations with images and animations, such as:

• Global patterns of currents

• Upwelling

• Black smokers

• Islands and atolls

DATA CENTERS

Extend your investigations with current and archived data and images, such as:

• Ocean currents

• Ocean floor

• Coral reefs

• Tides

EARTH SCIENCE NEWS

Relate your investigations to current events around world, such as:

• El Niño

• Sea level rise

• Undersea exploration

• Monsoons

Unit 6
Earth's Oceans

How Do Temperature and Salinity Affect Mixing in the Ocean? ▶ ES2202

1 Describe some of the general patterns you observe for temperature and salinity.

Higher salinity is generally found with warmer sea temperatures, especially those in the subtropics, and in secluded areas. Lower salinity is generally found with colder temperatures and along the continents, especially near river mouths.

2 Describe the location of at least three places in the world's oceans that have high temperatures but low salinity.

Answers will vary. Possibilities include the west coast of Africa at the equator, the north coast of Australia, the east coast of India, the coast around Southeast Asia, the northeast coast of South America, and along the west coast of Central America.

3 What do you think might cause low salinity in these areas of warm ocean water?

Answers may vary. Heavy rainfall or addition of fresh water from rivers might decrease salinity.

4 Determine the temperature, salinity, and density of ocean water at the following locations:
a. 0°N, 60°W   b. 20°N, 90°E

0°N, 60°W: Temperature = 27°C, Salinity = below 33‰, Density = 1.021 g/cm^3
20°N, 90°E: Temperature = 28°C, Salinity = below 33‰, Density = 1.021 g/cm^3

5 What is responsible for the unusually low salinity of these warm waters?

These locations are at the mouths of the Amazon and the Ganges rivers. They receive enormous amounts of fresh water, which lowers salinity.

6 Determine the temperature, salinity, and density of water on the Mediterranean and the Atlantic sides of the Strait of Gibraltar.

Mediterranean side: Temperature = 19°C, Salinity = above 37‰, Density = 1.027 g/cm^3
Atlantic side: Temperature = 19°C, Salinity = 36.4‰, Density = 1.026 g/cm^3

7 Predict what would happen to water that moves from the Mediterranean Sea into the Atlantic Ocean.

It would sink because it is more dense.

8 Determine the temperature, salinity, and density of the water on the Caribbean and the Pacific sides of the Panama Canal.

Caribbean side: Temperature = 28°C, Salinity = 35.8‰, Density = 1.023 g/cm^3
Pacific: Temperature = 28°C, Salinity = 33.6‰, Density = 1.021 g/cm^3

9 Predict what would happen to water that moves from the Caribbean Sea into the Pacific Ocean.

It would sink because it is more dense.

10 Find waters near Antarctica with the combination of the lowest sea temperatures and highest salinity. Extrapolate (extend the information on the graph) to determine the approximate density of these waters

Temperature = -1°C, Salinity = 34.4‰, Density = 1.028 g/cm^3

11 At what level in the ocean will these waters move away from Antarctica?

They would move along the bottom of the oceans, as they have very high density.

What's Responsible for Smaller Shrimp Catches?   ▶ ES2206

1 Examine the maps of mean oxygen levels throughout the year. Describe any patterns you see.

Answers will vary. Minimum oxygen values remain consistent throughout the year; maximum oxygen values drop dramatically in June.

2 Make two observations about the locations of "dead zones" in the world's oceans.

Answers will vary. Dead zones are mostly in the northern hemisphere, along coastlines of industrial countries, and near high-population centers.

3 Describe how the size of the Gulf dead zone changed from 1985 to 2000.

The area of the dead zone decreased from 1985-1988 and then generally increased until 1996. It jumped dramatically from 1992 to 1993. With the exception of 1999, when it was double its 1985 size, it seemed to shrink through the 1990's.

4 How does the area of the dead zone compare with nitrate flux over the same period? Describe the relationship.

The pattern of the size of the dead zone seems to be very similar to the nitrate flux for the same period.

5 Look at the maps of mean nitrate concentrations for four seasons. Describe your observations.

Nitrate concentration is relatively stable. Summer concentrations are similar to those in winter. Fall concentrations are identical to those in spring.

6 Look at the maps of mean chlorophyll concentrations for four seasons. Describe how concentrations change through the year.

There is an up-and-down trend: Very low in winter (0.2-0.4 mg/l), extremely high in spring (1.0-above 3.0 mg/l), very low again (lowest concentrations) in summer (0.05-0.1 mg/l), significantly higher levels in fall (0.2-3.0 mg/l).

7 Describe any relationship you observe between changes in oxygen levels and changes in nitrate and chlorophyll.

There doesn't seem to be a strong relationship between nitrates and the other factors. During the spring, phytoplankton blooms occur and oxygen decreases. Although phytoplankton production goes down in the summer, there is also a significant oxygen depletion then. Students may infer that there is a delay between phytoplankton production and oxygen depletion.

8 Describe the trend of catch per unit effort (CPUE) for brown shrimp from 1960 to 2000.

After 1981, the CPUE drops off dramatically. While fluctuations continue, generally CPUE continues to decline below CPUEs of the previous decades.

9 Do the CPUE data support the hypothesis that increased nitrate levels in the Mississippi River Basin reduce the shrimp harvests in the Gulf of Mexico? Why or why not?

The general downward trend of shrimp CPUE supports the hypothesis. CPUEs continue to fall while nitrogen use in the Mississippi River Basin continues.

10 Briefly explain how the geosphere, hydrosphere, and biosphere interact in this story.

Agriculture changes the geosphere in the Mississippi River basin with the addition of fertilizers to the soil. The hydrosphere runoff from fertilizers and organic forms of nitrogen from this farmland bring massive amounts of nitrates into the Gulf during the spring and summer. The introduction of these nitrates spurs an overabundance of phytoplankton. This phytoplankton bloom causes an increase in zooplankton and, as a result, an increase in waste products and decomposing algae. This, in turn, depletes oxygen levels. This may lead to lower or less efficient shrimp harvests as lack of oxygen drives off fish and shrimp.

Teacher's Guide Chapter 22
Internet Investigation

What Does the Ocean Floor Look Like? ▶ ES2301

1 What evidence did you look for to predict the location of each feature?

Answers will vary. The lowest sea depths indicate the deepest parts of the ocean where the ocean floor and trenches could exist. Shallower sea depths are closer to the surface and could indicate the presence of features such as ridges or seamounts. The shallowest waters would be over areas very close to the surface, such as the continental shelves.

2 Describe at least three observations from the animation.

Answers will vary. Water immediately drains off islands and the shelves that extend from the continent. It also drains early from the ridge in the middle of the ocean. Water drains from the continents toward the middle of the ocean and from the ridge toward the continents. Water drains last from the cracks (fracture zones) near the ridge. Even after all the water appears to have disappeared, a deep trench still holds some.

3 How do you think this ridge formed? Describe the geological process that occurs here.

Answers will vary. The ridge is a mid-ocean ridge—plates on either side of the ridge are diverging as new crust forms in the middle. Crustal material is relatively young, warm, and bouyant along the ridge.

4 Describe the sequence of how the water drains. Which features are exposed first? …last? Why does the water drain in this way?

Draining exposes the continental shelves and islands first, then the ridges. The water drains in toward the middle of the ocean along the basins and finally out of the fracture zones and trenches. It drains this way because shallowest areas empty first and deepest areas last.

5 How do you think this trench formed? Describe the geological processes that occur here.

The deep trench shows where one plate is being subducted under another plate.

6 Generally, where are passive and active continental margins located? How are these margins related to plate boundaries?

Almost all continental margins in the Atlantic Ocean are passive. Most continental margins in the Pacific are active; that is, they occur along plate boundaries.

7 Imagine that you could travel across the ocean floor from Lima, Peru to Shanghai, China. Narrate a brief "tour" of this path, describing ocean floor features you would encounter.

Answers will vary. Immediately after leaving the shelf, we drop into a trench, then climb up to an abyssal plain. At about 105W, 5S we ascend to a rise, which is a mid-ocean ridge. From here to about 150W, 0, we pass several fracture zones. We pass many islands and seamounts before arriving at another abyssal plain at 180, 10N. We pass around numerous islands on this plain. At about 145E, 15N there is a very deep trench. Coming out of the trench, we cross a ridge, a plain, and another ridge at 135E, 15N. We come to a deep trench at 130E, 25N. We pass a series of islands and come up the continental slope. At about 120E, 30N we climb the continental shelf and arrive at Shanghai.

Copyright © McDougal Littell Inc.

Teacher's Guide Chapter 23
Internet Investigation

Earth Science

When Were the Atlantic and Pacific Oceans Separated by Land?   ▶ ES2307

1 Make two observations about temperature and salinity on the Caribbean and Pacific sides of Panama.

The Caribbean side is warmer and saltier. The Pacific side is cooler and less salty.

2 Interpret the image to explain your observations.

Strong winds and plenty of sunshine evaporate Caribbean sea water , making it saltier. Trade winds, coming east from the Atlantic, carry evaporated water over to the Pacific side. This water condenses and provides abundant rainfall on the Pacific side. The abundance of fresh water, in the form of rain, lowers the salinity on the Pacific side.

3 Describe the changes in salinity in the Caribbean Sea and Pacific Ocean from six to two and a half million years ago.

The salinity of both sides was similar, and was relatively low from 6 to 5 mya. At about 4.8 mya, the salinity in the Caribbean and Pacific became very different. Salinity generally increased in the Caribbean and decreased in the Pacific.

4 Why do you think the ocean chemistry changed?

Answers will vary. The formation of a land bridge occurred. Isolated waters led to a change in climate that, in turn, caused changes in ocean chemistry.

5 How do you think this marine species could have spread to both sides of the isthmus?

At some time, the two bodies of water must have been connected, allowing these species to pass from one body to the other and exist on both sides.

6 Complete the chart below.

Sample #	Age (mya)	Pacific Present (Yes/No)	Caribbean Present (Yes/No)
1	6	No	No
2	5.6	No	No
3	5	No	No
4	4.8	Yes	No
5	4.3	Yes	No
6	3.5	Yes	No

7 How would you explain the presence of Pullentiatina only on the Pacific Ocean side?

This was a new species that developed when populations of microorganisms were isolated by the formation of a land bridge between North and South America.

8 How many years ago was the Isthmus of Panama formed? Explain how you used the data to determine this.

About 5 million years ago. This is about the time the new species, Pullentiatina, started to appear only in the Pacific, indicating that from that time after, it was cut off from the Caribbean. The graph showing the ratios between O-16 and O-18 also reveals a change occurred 5 mya.

9 Summarize the changes that have occurred in the Earth's spheres on either side of the Isthmus of Panama from six to three million years ago.

At 6 mya, the Caribbean and Pacific were connected. The salinity was very similar, and the same species of microorganisms were found on both sides. At about 4.8 mya, a land bridge was formed by the collision of the Nazca and Caribbean plates. This cut off the two bodies of water. From this time on, the salinity of water on both sides has diverged. The land bridge isolates populations of microorganisms, and a new species evolved on the Pacific side.

How Can One Ocean Current Affect the Whole North Atlantic?

1 Does it surprise you that palm trees could be growing in winter in Cork, Ireland? Why or why not?

Answers will vary. For example, Yes, because palm trees don't usually grow that far north

2 Record sea surface temperatures for January and July in your data table.

3 What patterns do you observe?

Answers will vary. The North American and inland European cities have colder average winter temperatures than coastal northern European cities. Coastal northern European cities have milder winters and a smaller range between average winter and summer temperatures.

4 Explain the effect of the Gulf Stream on the climate of coastal northern European cities.

The winter air temperatures of coastal northern European cities are warmed by the waters of the Gulf Stream. The current keeps temperatures fairly stable year-round.

5 Where are the two warm core eddies located in relation to the main Gulf Stream current?

The warm core eddies are located between 38N and 42N and 70W and 62W. They appear just north of the warmest band of the Gulf Stream.

6 What is the water temperature within the eddies? What is the temperature of the surrounding water?

The temperature of the water in the eddies is about 22 degrees Celsius. The water outside the eddies ranges from about 12 to 18 degrees Celsius.

7 Locate Gulf Stream eddies on this map. Are these warm or cold eddies? Briefly describe their location.

These are warm core eddies located north of the main Gulf Stream at about 37N-40N and 70W-65W.

8 Describe the concentration of phyto-plankton within these eddies. What can you infer about the abundance of fish in them?

The eddies have a very low concentration of phytoplankton, as does the main body of the Gulf Stream. Because of this low concentration, they are probably not productive fishing areas.

City, Country Longitude	Latitude,	Avg. high/low Jan. temps.	Avg. Jan. SST	Avg. high/low July temps.	Avg. July SST
Halifax, Canada	44N, 63W	0/-9° C	1° C	23/13° C	12° C
St. John's, Canada	45N, 66W	-2/-12° C	<0° C	21/12° C	8° C
Cork, Ireland	51N, 8W	7/2° C	8° C	19/10° C	14° C
London, UK	51N, 0W	7/2° C	6° C	21/13° C	14° C
Amsterdam, Neth.	52N, 4E	5/-6° C	4° C	21/15° C	14° C
Warsaw, Poland	52N, 21E	-1/-7° C	—	24/16° C	—
Kiev, Ukraine	50N, 30E	-3/-9° C	—	27/16° C	—

Teacher's Guide Chapter 24
Internet Investigation

1 What is the phase of the moon on October 17, 2001? On October 24?

New moon; First Quarter moon.

2 Predict the effect the moon's phase had on the level of the high tide and the low tide on October 17. Explain your reasoning.

Answers will vary. Very high high tides and very low low tidesoccurred because the sun's and moon's "pulling power" on Earth's water is combined in the same direction.

3 Predict the effect the moon's phase had on the level of the high tide and the low tide on October 24, Explain your reasoning.

Answers will vary. High and low tides would be at more moderate levels, because the sun and moon are at right angles to each other, which reduces the effect of the sun's gravitational pull on Earth's waters.

4 Describe the changes in tidal levels that occurred on October 17. At what times did the highest and lowest tides occur?

The tide is very low just after 7:00. The tide is coming in, because by 8:33, more of the shore is covered. By 10:35, the little channel is filled. By 12:20, almost all the land in the foreground is covered. This is the highest tide in the sequence. The tide goes out in such a way that 14:12 looks just like 9:35, 15:13 looks like 8:33, etc. By 18:10 all the land in the view is exposed again. This is the lowest tide.

5 There are roughly two high tides and two low tides every 24 hours. At what time do you think the next high tide occurred?

Answers will vary. The next high tide should be about 12 hours after the last high tide (or 6 hours after the last low tide), at about 12 a.m. (0:00) on October 18.

6 Describe the changes in tidal levels that occurred on October 24. At what times did the highest and lowest tides occur?

The morning images show the tide is going out. Highest tide in the morning sequence is in the first image at 8:44. The tide seems to be all the way out at 12:15, though not quite as far as the low tide on Oct 17. This is the lowest tide. From then, the tide comes in, and at 18:10 the water is at the highest point in the sequence.

7 Predict when the next low tide occurred.

The next low tide should occur about 12 hours after the last low tide (or 6 hours after the last high tide), at about 12 AM (0:00) on October 25.

8 Did the high tides occur at the same time on Oct. 17 and Oct. 24? Why or why not?

No, the high and low tides occurred around 6 hours (about 350 minutes) apart from Oct. 17 to Oct. 24. Because the tides correspond to the moon's cycle, which changes about 50 minutes each day, after seven days the high tide was 6 hours later.

9 Predict the changes in tide levels you would see at Cape Porpoise on October 31.

October 31, 2001 was a full moon, so tides would have been similar to those observed on October 17. High tide would have occurred around 12:00 (noon) and low tides at about 6:00 and 18:00.

Could You Break the Record for an Ocean Sailboat Race?

▶ ES2407

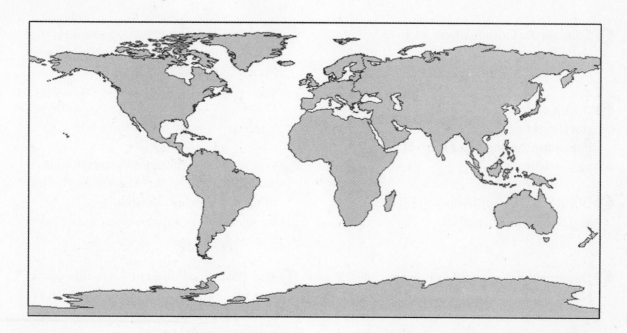

1 On the map, draw the route you would take to sail around the world's oceans.

2 Briefly explain your reasoning for selecting this route.

Answers will vary, but students should draw on their prior understanding of winds, ocean currents, climate, historic voyages, etc.

3 After viewing the Web pages, use a different color to draw a new or revised route. Explain in detail your reasoning for taking this route. Include notes and symbols on your map to help illustrate your explanation.

Answers will vary. Students should include an explanation of winds, sea currents, and climate conditions—those that are advantageous and those that pose dangers to be avoided.

For example: Leaving Boston Harbor, I would go in a southeasterly direction to take advantage of the prevailing winds. This would take me into the Gulf Stream at about 40°N, 65°W. This current and the

Westerlies would carry me across the Atlantic, north of the sluggish Sargasso Sea and the Horse Latitudes. At about 40°W, I would head SE toward the Canary Islands. At about 20° W, I would head south and slightly west toward the Cape Verde Islands. I would be helped by the North East Trades. By doing this, I would attempt to stay north of the Northern Equatorial Current, which is moving back west. Picking up the Canary Current at 20°W, 25°N, I would start heading for the Cape of Good Hope. I wouldn't want to pass too closely to Cape Verde or I might have to navigate coral reefs. Here, I would also have to monitor weather conditions for possible West African tornadoes. I would stay in the Canary current to get past the Doldrums at the equator. At the equator, I would go south and east, taking the Guinea Current along the coast of Africa. I would continue south along the African coast, staying east of the Benguela Current. From the Cape of Good Hope I would pick up the "Roaring 40's" and ride them SSE into the Indian Ocean. I hope I could stand the cold. I learned that sailboats don't have heaters, and latitudes south of 40°S are subject to winds from Antarctica, even in the southern summer… (continue).

Teacher's Guide Chapter 24
Internet Investigation

1 Estimate the latitude where warm El Niño water is located.

Near the equator, at 0° latitude.

2 During which months do ocean surface elevations begin to rise along the equator, indicating the beginning of an El Niño?

April – May, 1997

3 During which months does the El Niño end and the La Niña begin?

April – June 1998

4 Describe the thermocline during normal, El Niño, and La Niña phases.

Normal phase: 6,000 ft. deep in western Pacific; shallow (1,000 ft.) near South America. El Niño: 3,000 ft. deep in western Pacific; 5,000 ft. deep near South America. La Niña: about 3,000 ft across entire Pacific.

5 Describe changes in the ocean temperatures characterizing each phase.

Normal phase: warmest waters located in the western Pacific. El Niño: warmest water located in the eastern Pacific. La Niña: cooler and more uniformly distributed across the Pacific.

6 Choose one of the images or animations, and describe how it illustrates a particular part of the El Niño-La Niña cycle.

Answers will vary.

7 How does the North Pacific jet stream change from the mean during the January-March 1998 phase of the El Niño?

Instead of a continuous jet stream centered at 30°N latitude, it is broken into three shorter segments.

8 How does the North Pacific jet stream change from the mean during the January-March 1989 phase of the La Niña?

It becomes continuous, extending around the globe. Smaller jet stream segments are located in the middle of the Pacific Ocean.

9 Compare temperature and precipitation patterns around equatorial southeast Asia during El Niño and La Niña.

It is generally warm and dry in southeast Asia during El Niño and cooler and wet during La Niña.

10 How does the weather in the northwestern part of the United States differ between El Niño and La Niña? How does it differ in the southern United States?

Northwest U.S.: El Niño-warm; La Niña-cold and wet. Southern U.S.: El Niño-wet and cool; La Niña-dry and warm.

11 How have temperature and precipitation departed from the average for your area during El Niño conditions?

Answers will vary based on your location.

12 How have temperature and precipitation departed from the average for your area during La Niña conditions?

Answers will vary based on your location.

13 Describe an El Niño- or La Niña-related impact, and explain how El Niño or La Niña is thought to have caused this impact.

Answers will vary.

Can We Blame El Niño for Wild Weather?   ▶ ESU601

Unit 6 Project Description

How does El Niño-La Niña affect the Earth system? Use Web resources to identify, explore, and research an impact caused by El Niño or La Niña ocean circulation patterns. Develop a presentation that explains the impact and how it affects the Earth system. Present your findings in a format that has been approved by your teacher. Your project should include all components listed below.

Unit Project Criteria

1. **A description of the El Niño- or La Niña-related impact. (10%)**
   - Explain the impact, its location, and timing.
   - Use images, graphics, or animations to illustrate the impact.

2. **An explanation of the cause-and-effect relationship between El Niño-La Niña and the impact. (20%)**
   - Describe how El Niño-La Niña is thought to cause the impact.
   - Use images, graphics, or animations to illustrate the cause-and-effect relationship.

3. **An explanation of the scientific evidence for the cause-and-effect relationship. (25%)**
   - Interpret the scientific evidence, how it was gathered, and how it provides evidence supporting the cause-and-effect relationship.
   - Identify strengths and any uncertainties in the scientific evidence.

4. **A description of Earth system interactions involved in this cause-and-effect relationship. (25%)**
   - Identify and describe interactions in the Earth system related to the impact.
   - Illustrate interactions with images, graphics, or animations.

5. **A presentation of research and results. (20%)**
   - Communicate findings and ideas clearly and accurately.
   - Present appropriate data.
   - Use appealing visuals and design.

Earth System Interactions

Focus on the connections among Earth's spheres in your research and presentation. Think about how the impact you are researching affects our planet as a whole. For example, consider:

- El Niños and La Niñas are disruptions in normal ocean circulation (hydrosphere) patterns that affect worldwide weather (atmosphere), devastating crops and humans (biosphere) with droughts and flooding and causing abnormal erosion (geosphere).
- El Niño occurs when equatorial trade winds (atmosphere) decrease or reverse, which in turn causes changes in oceanic (hydrosphere) circulation patterns.
- Major shifts in ocean temperature (hydrosphere) cause shifts in the location and intensity of jet streams, which alter normal patterns of precipitation and temperature (atmosphere).
- Changes in weather (atmosphere) patterns result in changes to ecosystems (biosphere).

7 UNIT Space

NASA

Visit Mars from your desktop
Search for signs of water on Mars. Choose a landing site where you would look for signs of life.

Model asteroid impacts on Earth and the moon
Choose the size, speed, and density of asteroids that might collide with Earth and the moon. View examples of craters formed on each body.

Find how sunspots affect communications on Earth
Examine sunspots and explore how they not only cause beautiful auroras but also disrupt satellite communications on Earth.

Earth Science

7 UNIT   Space

OTHER WEB RESOURCES

VISUALIZATIONS

Spark new ideas for investigations with images and animations, such as:

• Solar System

• Impact theory of moon formation

• Radar mapping of Venus

• Life stages of stars

DATA CENTERS

Extend your investigations with current and archived data and images, such as:

• Mars exploration

• Remotely-accessed telescopes

• Night-time sky

• Meteorites

EARTH SCIENCE NEWS

Relate your investigations to current events around world, such as:

• Missions to Mars

• Search for extra-solar planets

• Solar and lunar eclipses

• New discoveries

Unit 7
Space

What if Earth and the Moon Were Hit by Twin Asteroids?   ▶ ES2506

1 Predict what would happen if twin asteroids, each 1000 meters in diameter, were to collide with Earth and the moon. Include a description of the impact craters you would expect to see on each body.

Answers will vary – look for an adequate description

2 What do you think the two craters would look like ten million years in the future?

Answers will vary. Lunar crater will change very little; Earth crater will be highly eroded or erased.

3 Calculate the "target area" of Earth and the moon. How many times larger a target is Earth than the moon? What does this say about Earth's chances of being hit by an asteroid compared to the moon's chances?

Earth = 127.8 million km^2; moon = 9.5 million km^2. Earth has about 13.5 times the cross-sectional area of the moon, making it much more likely to be hit by asteroids.

4 How do you think an ocean impact would differ from an impact on land?

Answers will vary. Ocean impacts could produce huge tsunamis that would flood coastlines across the globe. Land impacts would result in a crater, and huge amounts of rock materials would enter the atmosphere.

5 Describe the surface processes that work to erase craters and other impact features on Earth and the moon.

Earth's surface is changed by tectonic (mountain-building) processes, volcanism, erosion, and biological activity. The moon's surface is changed only by erosion from subsequent impacts.

6 The acceleration of gravity (g) on Earth is 9.8 meters/sec^2, but on the moon, it is only 1.6 meters/sec^2. How might this relate to the sizes of craters produced by impacts with identical amounts of kinetic energy on the two bodies?

Earth's higher g requires more energy for lifting materials from its surface to excavate a crater. Just as an astronaut can jump more easily on the moon, impact energy lifts material off the moon's surface more easily.

7 What happens to the crater diameter when you increase the speed of the asteroids?

Crater diameters increase.

8 What happens to the diameter of impact craters as the impact angle decreases?

Crater diameters decrease with decreased impact angle.

9 How does the kinetic energy of a 1000 m diameter iron asteroid traveling at 17 km/sec compare to the kinetic energy of an iron asteroid with twice the mass (diameter of 1260 km) traveling at the same speed?

Kinetic energy doubles with a doubling of mass.

10 How does the kinetic energy of a 1000 m diameter iron asteroid traveling at 17 km/sec compare to a similar one traveling at 34 km/sec?

Kinetic energy increases by a factor of four when the speed is doubled.

11 Use the Impact Calculator to determine the size and structure of craters on Earth and the moon produced by asteroids with 500 m diameters at different speeds, impact angles, and compositions. Make a table on a separate sheet of paper to record your results.

Students should show the estimated crater size and structure for a variety of 500 km diameter asteroids.

Earth Science

Why Does the Size of the Sun Appear to Change? ▶ ES2603

1 How would the sun appear in the animation if Earth orbited it in a perfect circle?

Apparent size of the sun would stay the same.

2 Sketch the shape of Earth's orbit that you think explains the sun's appearance in the animation.

Answers will vary. Should show some kind of oval or ellipse.

3 For each solar image, record the solar diameter, in pixels, in the table below.

Answers will vary slightly due to differences in measurement accuracy.

4 Calculate the distance to the sun for each month in millions of kilometers and record it in the table.

Date	Diameter (pixels)	Solar Distance (million km)
8/4/2000	467	152.1
9/4/2000	471	151.0
10/4/2000	475	149.5
11/3/2000	479	148.4
12/4/2000	483	147.2
1/4/2001	484	146.9
2/2/2001	483	147.1
3/5/2001	478	148.5
4/5/2001	474	149.8
5/5/2001	471	151.0
6/5/2001	468	151.8
7/4/2001	467	152.1

5 Describe the shape of the orbit you drew. How well does it match your prediction?

Answers will vary. Orbit should appear very nearly circular.

6 During which month does Earth's perihelion occur? When does Earth's aphelion occur?

Earth's perihelion occurs in January and aphelion occurs in July.

7 Which planet has the most circular orbit? Which planet has the least circular orbit?

Most circular – Venus (0.007); Least circular – Pluto (0.248)

8 What season is it in the northern hemisphere when Earth is closest to the sun (perihelion)?

Winter

How Does the Sunspot Cycle Affect Earth?   ▶ ES2605

1 Why do sunspots move across the face of the sun?

The sun is rotating.

2 What do the sunspot regions look like in the x-ray images? How might the activity at these areas affect Earth?

In the x-ray images, the sunspots look like fountains or explosions of x-ray energy. If directed toward Earth, energy and particles of the solar atmosphere may strike our atmosphere.

3 Earth is 150 million kilometers from the sun, and the solar "wind" travels at about 400 km/sec. How long does it take particles from a solar storm to reach Earth?

Around 104 hours or 4.3 days.

4 What do you think would happen if Earth's magnetic field became weaker or disappeared entirely?

Earth would no longer be well-protected from the solar wind. Any harmful solar effects would increase.

5 In the illustration, where do you see weaknesses in Earth's magnetic shield— places where the magnetic field may allow solar wind particles near the surface?

Near the poles.

6 Describe three ways that solar storms can affect your daily life.

Answers will vary. Examples: poor radio, TV, or cell phone reception, power outages, additional x-ray exposure in airplane flights, etc.

7 What is the average time interval between solar maxima?

11 peaks in 117 years = 10.6 years.

8 Predict the year in which you think the next solar maximum will occur.

Around 2008 (1997 + 11 years)

9 In the table, write the name of the environmental measurement and tell whether you think there is a correlation (connection) between the measurement and the sunspot number.

Answers will vary

Measurement	Correlation?
Ocean Temperature	Yes
Earthquakes	None
Precipitation	Maybe
Tree Rings	Maybe
Magnetic Disturbances	Yes, strong
Hurricanes	No

10 Identify three jobs held by people who would need to check the space weather report regularly, and explain why the report is important to each.

Answers will vary. Example: Pilot – navigation; Radio engineer – interference; Astronaut – radiation exposure.

How Fast Does the Wind Blow on Jupiter? ▶ ES2704

1 Why do we study the atmospheres of other planets in the solar system?

We do so to help us better understand Earth's atmosphere and the evolution of planetary atmospheres in general.

2 Describe the similarities and differences you observe between the global wind belts of Earth and Jupiter.

Answers will vary. Similarities: on both planets, northern and southern hemisphere belts are symmetrical and in alternating directions, and storms rotate in opposite directions. Differences: Jupiter's bands are more distinct and darker, with very little north-south cloud motion.

3 In the table, record the latitude, distance, and time of travel for each of the six features.

Latitude	Distance (km)	Time (hr)	Speed (km/hr)
30	96,390	780	124
-45	-14,053	780	-18
-35	35,252	780	45
-19	-146,933	780	-188
8	188,388	420	449
-37	31,274	780	40

4 Calculate the speed of each feature and record it in the table.

Example calculation:

96,390 km / 780 hr = 124 km/hr

5 The fastest wind speed ever recorded on Earth is 513 km/hr (318 miles per hour), measured during a May 1999 tornado in Oklahoma. How does the fastest wind speed you measured for Jupiter compare to this?

The fastest winds measured on Jupiter ≈ 450 km/hr (270 mph). Jupiter's "everyday" wind speeds are nearly as fast as Earths most intense storm winds.

6 How well do your data agree with the scientists' graph?

Answers will vary.

7 According to the graph, at what latitude are the fastest wind speeds on Jupiter?

around 30°N

8 Why do the winds on Jupiter blow faster from west to east than they do from east to west?

West-east winds blow in the direction of Jupiter's rotation; east-west winds blow against the planet's rotation.

9 What do you think are the differences between Earth and Jupiter that are responsible for the differences in their atmospheric circulation?

Answers will vary. The major differences are size, rotation rate, and surface features.

What Processes Shape Planetary Surfaces? ▶ ES2708

1 What was the first mission to explore a planet other than Earth or the moon? When did this occur, and which planet did it visit?

Mariner 2, 1962, Venus

2 How can a surface feature tell you about a surface process? Give an example.

Answers will vary. Example: A circular mountain with a central crater at the summit and evidence of flows on its flanks is probably a volcano.

3 List the processes that shape Earth's surface. Which of these surface processes do you think is the most common in the solar system? Which is the least common? Explain your answers.

Processes: Erosion (water, wind, ice), tectonics, volcanism, biological activity, and cratering.
Answers will vary. For example: cratering might be most common because space debris can hit any planetary body. Living things might be least common because Earth is the only planet known to support life.

4 What processes do you see evidence of on the surface of Mars?

Cratering, volcanism, erosion (water, wind, ice, landslides), tectonics (faulting)

5 What processes do you see evidence of on the surface of Venus?

Cratering, volcanism, tectonics (faulting)

6 What processes do you see evidence of on the surface of Mercury?

Cratering, tectonics (faulting)

7 What processes do you see evidence of on these moons? Identify the moon and the process you observe.

a) Earth's moon – cratering, volcanism; b) Io – volcanism; c) Triton – cratering, wind; d) Ganymede – cratering, faults, ice.

8 What is the most common surface process you observed in the solar system? Why do you think this process is so universal?

Cratering – the solar system is full of debris.

9 What processes have we seen in action on Mars? What process have we seen on Io?

Mars – wind and ice/frost, Io - volcanism

10 Which type of surface process appears to occur only on Earth? Why?

Biological processes; because Earth is the only planet known to support life.

11 How might humans shape the surface of other planets in the future?

Answers will vary. Examples: Constructing habitats, tapping geothermal resources, "terraforming", etc.

What Does the Spectrum of a Star Tell Us about Its Temperature? ▶ ES2803

1 How do you think the light from a star might indicate the star's temperature?

Answers will vary. Students may suggest that brighter stars are hotter or that different colors of light indicate different temperatures.

2 Record a list of the spectra in each of your groups. Describe the criteria you used to categorize the spectra.

Answers will vary. Students should list spectra assigned to each group and describe their categorizing scheme.

3 Record a list of the spectra you assigned to each group. How does this classification scheme compare with your first one?

Answers may vary. Class A: 2, 6, 7; Class B: 1, 3, 8; Class C: 5, 9, 12; Class D: 4, 10, 11.

4 Record the peak emission wavelength for each of the four standard stars.

A: 2800 Å; B: 3600 Å; C: 5500 Å; D: 7000 Å

5 Record the temperature indicated by peak emission wavelength for each of the four standard stars.

A: 10,000°C; B: 8,000°C; C: 5,000C; D: 4,000°C

6 Record the peak emission wavelength for each star's spectrum.

star 2: 2810 Å; star 6: 2812 Å; star 7: 2790 Å;
star 1: 3625 Å; star 3: 3612 Å; star 8: 3595 Å;
star 5: 5470 Å; star 9: 5510 Å; star 12: 5515 Å;
star 4: 7040 Å; star 10: 6940 Å; star 11: 7005 Å

7 Based on the peak emission wavelengths and the temperatures of the standard stars, estimate the temperatures of stars in each category.

Class A: 10,000°C; Class B: 8,000°C; Class C: 5,000°C; Class D: 4,000°C.

8 Based on their spectral patterns, estimate the temperatures of the four unknown stars.

W: 5,000°C; X: 8,000°C; Y: 10,000°C; Z: 4,000°C.

What Happens as a Star Runs Out of Hydrogen? ▶ ES2810

1 **What determines the size (volume) of a main sequence star?**

The size of a main sequence star depends on the initial mass of matter in the star and the balance between the inward force of gravity and outward force generated by fusion.

2 **Which size star uses up its available hydrogen most quickly? Explain your answer.**

The more massive the star, the hotter the interior becomes due to gravitational contraction. The hotter the core is, the more quickly its hydrogen fuel is converted to helium. The gravitational force of larger stars squeezes the hydrogen into a smaller volume than in smaller stars, so fusion reactions take place at a greater rate.

3 **After a massive star's hydrogen is depleted in the core, what does it use for nuclear fuel?**

Helium and carbon atoms fuse into progressively heavier elements, up to iron.

4 **Describe the final form of matter for large stars.**

The largest stars end as neutron stars—matter squeezed so tightly that electrons are forced into atomic nuclei and neutralize their protons, or as black holes—matter so dense that its gravitational field will not allow light to escape.

5 **Describe how a medium-sized star can grow as it runs out of hydrogen fuel.**

The extra energy given off when a stellar core begins fusing helium speeds up the hydrogen-to-helium nuclear reactions occurring in the shell. This causes higher outward pressure in the layers of gas above the core that isn't balanced by an increase in gravitational force, so the star's volume increases.

6 **How is the formation of a planetary nebula different from a supernova?**

Increased outward force from the core of a red giant creates a series of pulsations that gradually blow the outer shell of material away from the surface of the star. This outer shell of material sometimes appears as a "cloud" around the central star.

Supernova explosions eject the outer layers of a star in one incredibly violent explosion following the collapse of the iron core. The star is thought to implode, crush the core, then rebound. The supernova is the rebound. The amount of mass crashing down on the core determines the density of the star's final state.

7 **Describe the final form of matter for medium- and small-sized stars.**

Medium- and small-sized stars eventually become white dwarfs, stars that are extremely small (Earth-size) and very dense."

8 **How does the initial mass of a star compare to its final size (volume)?**

The larger (more massive) the initial star size, the smaller the final size. Large stars form black holes (with zero diameter) or neutron stars which are about as large as a city; medium and small stars form white dwarves that are Earth-sized objects.

Teacher's Guide Chapter 28
Internet Investigation

Earth Science

Could Mars Support Life?

1 List at least two factors that make life possible on Earth.

Answers will vary. Students may mention the presence of liquid water, an atmosphere that shields Earth from excessive ultraviolet radiation and moderates temperature, and the availability of light energy or chemical energy.

2 List at least two similarities and two differences between Earth and Mars.

Answers will vary. Both are rocky planets that have white ice caps. The diameter of Mars is about half that of Earth. No oceans or clouds are apparent on Mars.

3 What do these images tell you about the two planets?

The images provide evidence of atmospheric circulation. Some students may notice that both storms radiate from a central vortex. The presence of wind and atmospheric dust suggests that both planets experience erosion.

4 List at least two similarities and two differences in the ground-based views of Earth and Mars.

Answers will vary. Both images show a range of rock sizes, and both lack vegetation. Mars shows no water or clouds. The Martian atmosphere is slightly pink rather than blue.

5 In which of these habitats are you most surprised to find life, and why?

Answers will vary. Look for justification of whichever habitat students choose.

6 What similarities do you see between the features on Mars and those formed by flowing water on Earth?

Answers will vary. Students may note presence of canyons, channels, valleys, or layers of sediment in the Mars image.

7 Record your observations for each site.

Mars Image	Evidence of water?	Description of evidence
Site 1	Yes, flowing	Presence of grainy material; crater channels have been breached
Site 2	Yes, flowing	Presence of rampart craters; a flood plain is evident
Site 3	Yes, flowing	Presence of water-deposited sediments
Site 4	Possible source	This volcano could serve as a heat source to melt ice.
Site 5	Yes, flowing	Presence of grainy surface; a flood plain is evident
Site 6	Possible source	Polar cap is made of ice that, given a heat source, could melt
Site 7	Yes, flowing	Presence of gullies along the crater wall
Site 8	Yes, flowing	Presence of water-deposited sediments

8 If you were to lead a mission to Mars, at which of the sites would you choose to land in your search for life on Mars? Describe why you think this site has the greatest chance of exhibiting evidence of life.

Answers will vary.

Teacher's Guide Unit 7
Internet Investigation

Could Mars Support Life?

Unit 7 Project Description

Using the results of this investigation and Web-based research on current missions to Mars, prepare a presentation to argue a case either for or against the likelihood that Mars has supported, or could support, life. The case you make must be based on data and research. Present your findings in a format that has been approved by your teacher. Your project should include all components listed below.

Unit Project Criteria

1. **A description of the requirements for life. (25%)**
   - Describe the requirements for life on Earth.
   - Explain what life in extreme habitats on Earth tells you about the possibility of life on Mars.
   - Describe similarities and differences between Mars and Earth related to their abilities to support life.
2. **An evaluation of the data from the eight sites on Mars. (40%)**
   - Summarize your observations.
   - Evaluate the evidence of flowing water or other signs related to the potential for life.
3. **An explanation of the findings of your Web-based research. (20%)**
   - Summarize your findings from recent missions to Mars.
   - Use images, graphics, or animations to illustrate the findings.
4. **An effective presentation of your conclusions. (15%)**
   - Communicate findings and ideas clearly and accurately.
   - Present appropriate data that support your conclusions.
   - Use appealing visuals and design.

Earth System Interactions

Focus on similarities between the Mars system and the Earth system in your research and presentation. For example, consider:

- Living organisms (biosphere) on Earth rely on water (hydrosphere) for their existence.
- Flowing water (hydrosphere) shapes rocky surfaces (geosphere).
- The atmosphere shields living organisms (biosphere) from harmful radiation and moderates surface temperatures.
- Water (hydrosphere) can be held in the atmosphere or in polar ice caps or groundwater (geosphere).

8 UNIT Earth's History

Jennifer Loomis, TERC

Explore the Grand Canyon
Read the rock record of environmental change revealed by layers of the Grand Canyon.

Use fossils to investigate how life has changed
Compare fossils from different periods of Earth's history to explore how life has changed over time.

Investigate the causes of mass extinctions
Explore evidence from the rock record to look for the causes of mass extinction events.

Unit 8
Earth's History

UNIT Earth's History

OTHER WEB RESOURCES

VISUALIZATIONS

Spark new ideas for investigations with images and animations, such as:

• Events thru geologic time

• How fossils form

• Break-up of Pangea

• Asteroid impacts

DATA CENTERS

Extend your investigations with current and archived data and images, such as:

• Determining the age of rocks

• Grand Canyon case study

• Paleontology

• Reading history in sediments

EARTH SCIENCE NEWS

Relate your investigations to current events around world, such as:

• Fossil finds

• Endangered species

• New discoveries and insights

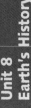

Unit 8
Earth's History

Copyright © McDougal Littell Inc.

❶ Identify at least three geologic processes represented in the images.

Answers will vary. Students may mention volcanism, erosion, deposition, and faulting.

❷ What evidence might indicate that these processes also have occurred sometime in the past?

Answers will vary. Possibilities include the presence of layers of basalt, layers of sediments, offset layers, and missing layers.

❸ Which layers were deposited first and are therefore the oldest?

The layers at the bottom were deposited first and therefore are the oldest.

❹ Which layers were deposited last and are therefore the youngest?

The layers at the top were deposited last and therefore are the youngest.

❺ How might layers that were once horizontal come to be tilted like this?

Regional uplifting or igneous intrusions can cause originally horizontal layers to become tilted.

❻ What type of force was applied to these rocks to form folds?

These layers were pushed together by compression.

❼ Identify and label the unconformity on your diagram. Number the layers from oldest to youngest.

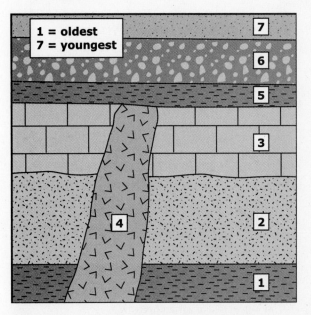

1 = oldest
7 = youngest

❽ Make a sketch of the layers in the image. Then, number the layers from oldest to youngest. Next, use another sheet of paper to write a detailed description of the geologic history of the site.

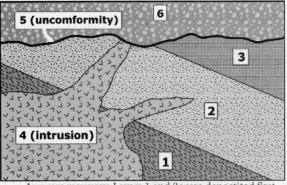

5 (uncomformity)
4 (intrusion)

Answers may vary. Layers 1 and 2 were depostited first. They were originally horizontal. The relative ages of layer 3 and the intrusion at 4 cannot be determined with certainty. Layers 1, 2, and 3 were tilted, perhaps by the geologic event that formed the intrusion. The entire region experienced uplift and erosion, followed by futher deposition of layer 6.

Teacher's Guide Chapter 29
Internet Investigation

1 Approximately how old is this tree? Where are the oldest rings?

The cross section has approximately 35 rings, indicating the tree is 35 years old. The central rings are the oldest. They represent the tree's first years of growth.

2 Which rings indicate years of relatively abundant water? Which rings might indicate years of drought?

Rings 1, 2, and 4 indicate years of abundant water. Rings 3 and 5 indicate limited water availability.

3 Other than forest fires, what else might tree rings be used to date?

Answers will vary. Tree rings might be used to date buildings or fence posts. Other environmental uses include dating floods, insect damage, glacial movements, and volcanic activity.

4 How old is this tree?

Core 1 is from a tree that is 20 years old.

5 How old is this tree?

Core 2 is from a tree that is 20 years old.

6 How do the banding patterns of the two cores compare?

The banding patterns of both cores reflect the same wet and dry years. The width of rings in the tree on the bottom is twice that of the tree on top.

7 Transfer the plot onto your graph paper.

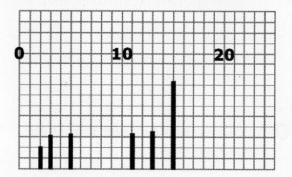

8 How do the skeleton plots compare?

The skeleton plots of these cores look identical.

9 Compile the master chronology on the graph paper at the bottom of your worksheet.

10 What year was tree sample A cut?

Tree sample A was cut in 1091.

11 What year did tree sample B begin growing?

Tree sample B began growing in 1028.

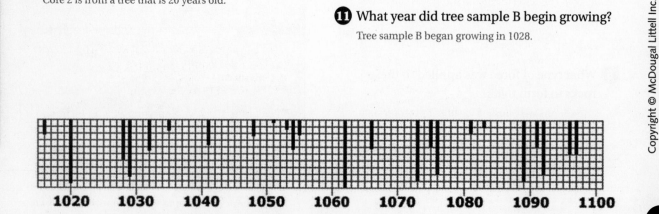

How Did the Layers of the Grand Canyon Form? ▶ ES2906

1 What are some of the ways the sedimentary rock layers may have formed?

Answers will vary. Students may mention sediments accumulating in an ocean basin, sand dunes covering the area, rivers filling in a low valley.

2 When did the rock layers exposed in the canyon form?

Began approximately 2 billion years ago and ended about 250 million years ago.

3 Are there any major gaps in the rock record? If so, when?

Yes. The Silurian period, between 430 and 395 million years ago, is missing

4 The Zoroaster granite is statigraphically lower than the Vishnu schist, but it is younger than the schist. Explain how this relationship is possible.

The granite intruded into the previously existing schist. (The principle of superposition only holds true for sedimentary layers that have not been overturned.)

5 Describe the events necessary to produce the observed rock sequence of horizontal layers on top of the tilted layers of the Grand Canyon Supergroup.

Folding or faulting of the originally horizontal layers tilted the Supergroup. Erosion of the tilted layers occurred before the next horizontal layers were deposited on top of them.

6 Is it possible to estimate how much material was eroded away? Explain your answer.

No. As the eroded material is not present, there is no way to tell if one foot or 1000 feet of material were eroded.

7 Describe the changing environmental conditions that resulted in the formation of the three layers of the Tonto Group.

An advancing sea is indicated by the successive layers—first a beach, then a shallow sea, then a deeper sea covered this area.

8 What geological processes could account for the unconformity between the Mauv and the Temple Butte Layers?

Lack of deposition or uplift and erosion.

9 Describe the rock type and depositional environment of the two similar layers; then discuss how they differ from the third.

The Temple Butte and Redwall layers are both limestones that were formed in shallow seas. The Surprise Canyon formation reflects a change in the environment toward a shallower sea—a tidal estuary on a low coastal plain.

10 Describe the environment in which the Supai Group and Hermit Shale were deposited. Describe the fossils that are associated with these layers.

Both were deposited in swampy environments, intermingled with deltas and coastal dunes. The fossils from these environments include amphibian and reptile tracks and the impressions of ferns and trees.

11 Based on the rocks, how might the climate of the area have changed between deposition of the Hermit Shale and the Coconino Sandstone?

The climate appears to have become drier. Windblown sandstone indicates a desert environment; the shale was deposited in a muddy environment .

12 Hypothesize about why there are no layers younger than 225 million years old in the Grand Canyon area.

Answers will vary. The younger layers have all been eroded away, or they were never deposited in the area.

How Has Life Changed over Geologic Time? ▶ ES3002

1 Write down the name of at least one of the fossils that you think you can identify.

Answers will vary. Accept all responses.

2 What fossils have been found in your state?

Answers will vary. Accept all responses.

3 What type of fossils would you expect to find in an environment with sand dunes?

You might expect to find the fossilized remains of desert dwelling organisms, such as lizards and drought-adapted plants.

4 For each site on your worksheet, record the kinds of fossils that are found, the environment in which they lived, and the time frame during which they lived. It may also be helpful to record any geologic periods or epochs that are mentioned.

5 How has life changed over geologic time? Describe the patterns that you see.

Life originated in the sea and eventually spread out onto land. The progression in the fossil record shows marine organisms, followed by amphibians, followed by land creatures. In the history of life on Earth, simple organisms arose before more complex ones.

6 Make a sketch of your favorite fossil.

Answers will vary. Accept all responses.

7 Write a one-paragraph description to accompany your sketch.

Answers will vary.

Site	Fossils Present	Environmental Conditions	Time Span
LaBrea Tar Pits	Sabre-tooth cat, mammoth, gray wolf	Abundant tar pits containing asphalt in sticky pools	28,000 years ago Pleistocene
Hagerman Fossil Beds	Hagerman horse, beaver, muskrat, camel, ground sloth	Grassy plains, ponds, and forests	3.7 *mya* Pliocene
Agate Fossil Beds	Menoceras (2-horned rhino) Moropus (horse-like organism) Dinohyus (huge, pig-like creature)	Widespread drought occurring in a savanna-like habitat	19 *mya* Miocene
Dinosaur National Monument	Stegosaurus, Allosaurus, Apatosaurus	A large plain crossed by several rivers and streams with associated sandy and gravelly banks; Ferns, cycads, and conifers were nearby	145 *mya* Jurassic
Petrified Forest	Coniferous trees	A large basin with rivers, streams and conifers over 150 feet tall.	225 *mya* Triassic
Guadalupe Mountains	Dimetrodon	Capitan reef of the Delaware Sea and nearby coastal region	260 *mya* Permian
Mazon Creek Fossils	Shrimp, fern, coal	Swamps and shallow bays with abundant deltas	300 *mya* Pennsylvanian
Ohio River Fossil Beds	Fish, brachiopods	Warm shallow seas	386 *mya* Devonian
Burgess Shale	Trilobites, sponges, worms, and other marine invertebrates	Warm shallow seas	540 *mya* Cambrian

Teacher's Guide Chapter 30 Internet Investigation

Copyright © McDougal Littell Inc.

Where and When Did Dinosaurs Live? ► ES3008

1 Which of these dinosaurs do you recognize or know something about?
Accept all answers.

2 Write down the name of one of these dinosaurs that you would like to know more about.
Accept all answers.

3 Prepare a case study of a day in the life of your dinosaur. Be sure to include the following:
- the time frame in which your dinosaur lived
- the type of food it ate
- its closest relatives
- the most common places to find its fossils
- the environmental setting during the time it lived
- at least three characteristics that make this dinosaur unique

4 Share your report with others and compile the information about other dinosaurs into the table.

5 Could a *Tyrannasaurus Rex* have eaten a Stegosaurus for lunch? Why or why not?
No. They were separated in time by about 80 million years.

6 On a separate sheet of paper, draw a timeline for the dinosaurs in the table.
Check to see that the order of dinosaurs on the timeline matches the one given in the table below.

Geologic Era	Dinosaur Name	Time Span	Food Eaten	Fossil Locations	Closest Relatives
Cretaceous	Triceratops "three-horn face"	65 mya	Low growing plants	Western U.S., Southwest Canada	Centrosaurus, a one-horned dino, "pointed lizard"
	Tyrannosaurus "tyrant lizard"	67 mya	Meat eater, perhaps triceratops	North America, Central Asia	Albertosaurus "Alberta lizard"
	Velociraptor "swift robber"	67 mya	Meat eater	Central Asia	Deinonychus "terrible claw"
	Ankylosaurus "stiff joint lizard"	70 mya	Soft vegetation	Western U.S.	Euplocephalus
	Pteranodon "wing without tooth", not a true dinosaur	85 mya	Fish, perhaps a scavenger as well	Central U.S.	--------------
Jurassic	Allosaurus "different lizard"	150 mya	Meat eater, likely large animals	Western U.S.	Ceratosaurus "horned lizard"
	Camarasaurus "chamber lizard"	152 mya	Plant eater	Western U.S., Northwestern Mexico	Brachiosaurus "arm lizard"
	Apatosaurus "deceptive lizard"	152 mya	Plant eater	Western U.S. Northwestern Mexico	Diplodocus "double beam"
	Diplodocus "double beam"	152 mya	Plants low to the ground	Western U.S.	Apatosaurus "deceptive lizard"
	Stegosaurus "roofed lizard"	155 mya	Low-growing ferns	Western U.S.	Toujiangosaurus

What Caused the Mass Extinction Recorded at the K-T Boundary?

▶ ESU801

1 What biological findings have been reported at the Cretaceous-Tertiary (K-T) boundary?

The fossil record indicates that more than half of all plants and animals living on Earth became extinct. These include the dinosaurs, the pterosaurs, many species of plants, and marine organisms such as the ammonites, marine reptiles, and several species of bivalves.

2 What geological findings have been reported at the K-T boundary?

The K-T layer is characterized by the presence of iridium, tektites, and shocked quartz.

3 What is significant about the discovery of iridium at the K-T boundary?

The discovery of unusually high levels of iridium suggests an extraterrestrial origin of the clay layer, perhaps caused by an asteroid or a comet.

4 What are the Deccan traps, and what theory do they help support?

The Deccan traps are large basalt flows that flooded India 65 million years ago. They provide evidence that intense volcanism may have contributed to widespread environmental changes at the K-T boundary.

5 How has discovery of the Chicxulub impact crater shifted the debate?

Most researchers now accept that an impact occurred 65 million years ago. Current research is still investigating the mechanism by which such an impact could have induced the types of losses recorded in the fossil record.

6 Describe how Earth's systems might interact after an asteroid impact.

Answers will vary. Thousands of tons of debris were blasted into the atmosphere, producing dust that blocked out sunlight. Reduced sunlight and lowered temperatures in the atmosphere affected the biosphere, limiting the abilities of plants to grow and limiting food supplies for animals. At the point of impact, the geosphere experienced deformation, shockwaves, and earthquakes, all of which affected the biosphere.

7 Write at least one question about mass extinction that this investigation has raised for you.

Answers will vary. Examples include: How do the frequency and timing of impact cratering events compare to the frequency and timing of mass extinctions? How does the global distribution of impact craters compare to the global distribution of basalt flows?

8 How might you investigate this question?

Answers will vary.

Teacher's Guide Unit 8
Internet Investigation

What Caused the Mass Extinction Recorded at the K-T Boundary?

Unit 8 Project Description

Using the results of this investigation and Web-based research on current and new discoveries about mass extinctions, prepare a presentation that builds a case to support an existing theory, or that formulates a new theory, explaining the cause of a mass extinction. The case you make must be based on data and research. Present your findings in a format that has been approved by your teacher. Your project should include all components listed below.

Unit Project Criteria

1. **A description of the location where the extinction event occurred. (10%)**
   - Prepare a map showing specific location(s) involved with the extinction event.
   - Cite evidence to support the location(s) you identified for the extinction event.

2. **An explanation of when the extinction event occurred. (10%)**
   - Tell how many years ago and in which evolutionary period the event occurred.
   - Cite evidence that allows you to date the event.

3. **An explanation of the evidence you used to make your case. (20%)**
   - Explain the source from which your evidence comes (geological, biological, extraterrestrial).
   - Explain clearly how you interpret the evidence.

4. **A summary of your theory of how the extinction occurred. (40%)**
   - Explain how the evidence supports this.
   - Explain the story in terms of interactions among Earth's spheres.

5. **An effective presentation of your conclusions. (20%)**
   - Communicate findings and ideas clearly and accurately.
   - Present appropriate data that support your conclusions.
   - Use appealing visuals and design.

Earth System Interactions

Focus on the connections among Earth's spheres in your research and presentation. Think about the interactions among spheres that resulted in a mass extinction of the biosphere. For example, consider:

- An asteroid impact would have affected all of Earth's spheres in a synergistic manner.
- Thousands of tons of debris were released into the atmosphere, producing dust that encircled the globe and blocked out sunlight.
- Reduced sunlight and lowered temperatures in the atmosphere affected the biosphere, limiting the ability of plants to grow and limiting food supplies for animals.
- At the point of impact, the geosphere experienced deformation, shockwaves, and earthquakes, all of which affected the biosphere.
- Tidal waves and hurricanes in the hydrosphere affected the geosphere and the biosphere.
- Acid rain in the atmosphere damaged the biosphere.